Scientific Methods
of Inquiry for
Intelligence Analysis

SECURITY AND PROFESSIONAL INTELLIGENCE EDUCATION SERIES (SPIES)

Editor: Jan Goldman

In this post–September 11, 2001 era, there has been rapid growth in the number of professional intelligence training and educational programs across the United States and abroad. Colleges and universities, as well as high schools, are developing programs and courses in homeland security, intelligence analysis, and law enforcement, in support of national security. The Security and Professional Intelligence Education Series (SPIES) was first designed for individuals studying for careers in intelligence and to help improve the skills of those already in the profession; however, it was also developed to educate the public in how intelligence work is conducted and should be conducted in this important and vital profession.

Books in the series include:

Communicating with Intelligence: Writing and Briefing in the Intelligence and National Security Communities, by James S. Major, 2008

A Spy's Résumé: Confessions of a Maverick Intelligence Professional and Misadventure Capitalist, by Marc Anthony Viola, 2008

An Introduction to Intelligence Research and Analysis, by Jerome Clauser, revised and edited by Jan Goldman, 2008

Writing Classified and Unclassified Papers for National Security, by James S. Major, 2009

Strategic Intelligence: A Handbook for Practitioners, Managers, and Users, revised edition by Don McDowell, 2009

Partly Cloudy: Ethics in War, Espionage, Covert Action, and Interrogation, by David L. Perry, 2009

Ethics of Spying: A Reader for the Intelligence Professional, edited by Jan Goldman, 2006 (volume 1) and 2010 (volume 2)

Handbook of Warning Intelligence: Assessing the Threat to National Security, by Cynthia Grabo, 2010

Handbook of Scientific Methods of Inquiry for Intelligence Analysis, by Hank Prunckun, 2010

Keeping U.S. Intelligence Effective: The Need for a Revolution in Intelligence Affairs, by William J. Lahneman, 2011

Words of Intelligence: An Intelligence Professional's Lexicon for Domestic and Foreign Threats, second edition, by Jan Goldman, 2011

Balancing Liberty and Security: An Ethical Study of U.S. Foreign Intelligence Surveillance, 2001-2009, by Michelle Louise Atkin, 2013

Communicating with Intelligence: Writing and Briefing in National Security, second edition, by James S. Major, 2014

Quantitative Intelligence Analysis: Applied Analytic Models, Simulations and Games, by Edward Waltz, 2014

The Art of Intelligence: Simulations, Exercises, and Games, edited by William J. Lahneman and Rubén Arcos, 2014

Scientific Methods of Inquiry for Intelligence Analysis

Second Edition

Hank Prunckun

ROWMAN & LITTLEFIELD
Lanham • Boulder • New York • London

Published by Rowman & Littlefield
A wholly owned subsidiary of The Rowman & Littlefield Publishing Group, Inc.
4501 Forbes Boulevard, Suite 200, Lanham, Maryland 20706
www.rowman.com

16 Carlisle Street, London W1D 3BT, United Kingdom

British Library Cataloguing in Publication Information Available

Library of Congress Cataloging-in-Publication Data

Prunckun, Hank, 1954–
  [Handbook of scientific methods of inquiry for intelligence analysis]
  Scientific methods of inquiry for intelligence analysis / Hank
Prunckun.—Second Edition.
    pages cm.— (Security and professional intelligence education series ; 19)
  Revised edition of: Handbook of scientific methods of inquiry for intelligence analysis.
  Includes bibliographical references and index.
  ISBN 978-1-4422-2431-5 (cloth : alk. paper)—ISBN 978-1-4422-2432-2 (pbk. : alk. paper)—ISBN 978-1-4422-2433-9 (electronic) 1. Intelligence service—Methodology—Handbooks, manuals, etc. 2. Science—Methodology—Handbooks, manuals, etc. 3. Social sciences—Methodology—Handbooks, manuals, etc. 4. Behavioral assessment—Methodology—Handbooks, manuals, etc. I. Title.
JF1525.I6P78   2014
327.12072—dc23

                                                                 2014018227

♾ ™ The paper used in this publication meets the minimum requirements of American National Standard for Information Sciences—Permanence of Paper for Printed Library Materials, ANSI/NISO Z39.48-1992.

Printed in the United States of America

Because of the nature of intelligence work, there are cadres of scholar-spies who will never be recognized for their contribution to the safety and well-being of society. However, I acknowledge your endeavors and I dedicate this book to you.

Contents

Series Editor Foreword

A decade ago the Security and Professional Intelligence Education Series (SPIES) began offering publications. I have been extremely careful to choose books that further develop the emerging intelligence profession. Unlike other publications that "talk about intelligence," this would be the first series to focus on "doing" intelligence.

This is the second edition of a book that has been well received in the intelligence analytical and academic communities. Many of the chapters have been revised to include new material, detailed explanations, study questions, and additional illustrations. New sections can be found in almost every chapter. For example, in the chapter on idea generation and conceptualization there are new sections on random input, sorting, and many-to-many matrix. This edition also includes four new chapters.

Slowly the field of knowledge is evolving and expanding for the professional intelligence analyst, and this book and SPIES are representative of this trend. I am proud to say this is the first time we have had the opportunity to offer a revised edition of a popular book, and I hope we can do this for other books in the series when a new edition is warranted.

Jan Goldman
Washington, DC

Preface

In the first edition of this book I argued that no other profession has experienced change to the same extent that intelligence has since the terrorist attacks of September 11, 2001. At the writing of this second edition, I am of the same view. Intelligence has grown much larger, and its mission is more complex. This has been evidenced by government and private sector security agencies recruiting intelligence analysts to process what has become a voluminous amount of raw information that has flowed into these agencies' data-collection systems.

Despite the demand for analysts to process these data into intelligence products, the reality has been that there has been an unmet need for trained analysts. For this reason, we have witnessed a growing number of colleges and universities offering intelligence study programs so candidates for analyst positions can begin their duties without protracted on-the-job instruction. This second edition is not just a summary of existing knowledge; it provides results of an investigation into the essential analytic skills critical for undertaking intelligence work. In this regard, this new edition provides the theoretical foundations as well as practical insights into the science of intelligence. Therefore, it is equally useful for researchers and "scholar-spies."

This edition has been revised and expanded to address the subject of intelligence analysis with a grounding in what I argue are the origins of intelligence research—the scientific method of inquiry at the core of such academic disciplines as sociology, anthropology, criminology, psychology, political science, history, economics, education, and library science. Although there are many intelligence texts covering aspects of what is discussed in these pages, in my view this second edition is a more comprehensive text because it covers the topic in a holistic way—it is more than a textbook on analytic techniques—it is a serious piece of research on the theory and practice of intelligence analysis.

While the literature on intelligence abounds with works on spy gadgetry and covert surveillance, one has to look wider for material on intelligence research and analysis. Although an increasing number of texts on analytic techniques are being published, this book goes further—it not only presents the reader with a range of analytical methods used in secret research, but it explains how secret intelligence fits into the larger research framework.

This second edition discusses not only the essentials of applied intelligence research but also analyzes the function, structure, and operational methods involved in intelligence work. In particular, it explores how an analyst will be required to obtain data via covert methods, and then work with this classified information in a security-conscious environment. It also examines how intelligence data are validated, in marked contrast to how a social science researcher performs the same task. The reader is left with little doubt about the theoretical foundations of intelligence, how intelligence is developed, and how it is processed in an environment that has security and secrecy at its core.

The need for such a book was born out of my personal experience as a researcher and analyst. In many of the positions I held during my career, I relied on texts in other academic disciplines, as none addressed the science of intelligence. Occasionally I found texts in the field of criminal justice and police science that were of value, but again they addressed issues faced by analysts obliquely. There are several excellent texts for industry-specific applications—for instance, national security intelligence, military intelligence, or law enforcement intelligence—but by definition these texts are narrow in focus (e.g., foreign policy- or police-centric) and do not apply the principles of intelligence holistically across the spectrum of industries that ply this craft.

The second edition of *Scientific Methods of Inquiry for Intelligence Analysis* examines how these concepts can be applied in the post-9/11 world—an environment that has less-clear boundaries between what may have been military intelligence and, say, business intelligence, and so forth. This edition examines how applied research methods are used by intelligence practitioners to conduct the secret work they do. It is a systematic exploration of the theoretical concepts within the intelligence discipline, thereby providing scholars and practitioners with the knowledge of how to be effective researchers in a variety of intelligence settings: military, national security, law enforcement, business, and the private sector.

The book comprises twenty-two unique topics. Each topic contains a number of concepts that build into a thorough understanding of intelligence research and analysis. This edition includes new chapters based on the research I conducted since the release of the first edition. This new material addresses topics such as open sources of information; content analysis of qualitative data; target profiles; tactical assessments; vehicle route analysis; and decision analysis, as well as others. In addition to these new research findings, all other chapters have undergone revision to ensure that their content reflects the new developments and thinking relevant to the craft of intelligence analysis.

Hank Prunckun, PhD
Sydney, 2014

1

⑤

Intelligence Theory

This topic provides an introduction to intelligence research by examining:

1. Intelligence research;
2. Why intelligence;
3. Information versus intelligence;
4. Intelligence defined;
5. Intelligence as knowledge;
6. Intelligence as a process;
7. Intelligence versus investigation;
8. Data versus information; and
9. Intelligence theory.

INTELLIGENCE RESEARCH—
A HARD ROW TO HOE

It could only be described as a typical winter's day in late January 1993. Like most days, a line of traffic came to a halt at a set of traffic lights on the eastbound lane of Virginia Route 123. This place was just outside the entrance to CIA headquarters in Fairfax County, Virginia. In the line of cars waiting to go to work were CIA staffers and various contractors. It was a routine day for them as well as the many intelligence analysts who were already at work within the complex.

On such a day, one would not expect analysts to face anything more dangerous than the hazard posed by some careless driver. But intelligence work—no matter how remote analysts are from the James Bond–like scenarios of intelligence gathering—is a profession fraught with danger.

1

On January 23, 1993, a Pakistani assassin[1] in that same line of traffic got out of his car at the stoplight and walked calmly from vehicle to vehicle shooting the male occupants with his AK-47 assault rifle. He only stopped firing because, as he later confessed, he ran out of targets. Among those wounded and dead were intelligence analysts.

Although this book is a critical discussion of intelligence research—a seemingly urbane profession—make no mistake: intelligence work carries with it dangers. The CIA Memorial Wall (and Book of Honor, see figure 1.1) displays stars that represent those who gave their lives for their country in the service of intelligence.[2] And yes, among these people are analysts who conducted the genteel craft of intelligence research. It is a "hard row to hoe," both mentally and physically. As the former CIA director Robert Gates once stated: "The nation is at peace because we in intelligence are constantly at war."[3]

WHY INTELLIGENCE

Why be concerned with intelligence? Because intelligence enables one to exercise control over a given situation. In this sense, control equates to power. Ira Cohen, in his classic treatment of the study of power, wrote:

> Power is sought because without power the security and even the ability of [one] to continue to exist is generally decreased. Without power, [one] has no ability to deter another . . . from actions whose consequences threaten the vital interests of the former. Without power [one] cannot cause another . . . to do that which the former desires but which the latter desires not to do. Power is sought because the more power that [one] has, the greater is the number of [his or her] available options. The more options available to [one], the greater [his or her] security. The greater [his or her] security, the better off [he or she is]. [He or she is] more secure in [his or her] life and in the enjoyment of [his or her] private property.[4]

Intelligence is, therefore, not a form of clairvoyance used to predict the future but an exact science based on sound quantitative and qualitative research methods. But as Lowenthal points out: "Intelligence is not about truth. If something were known to be true, states would not need intelligence agencies to collect the information or analyze it. . . . [So,] we should think of intelligence as a proximate reality. . . . [Intelligence agencies] can rarely be assured that even their best and most considered analysis is true. Their goals are intelligence products that are reliable, unbiased, and honest (that is, free from politicization)."[5] In this regard, intelligence enables the analyst to present solutions or options to decision makers based on defensible conclusions.

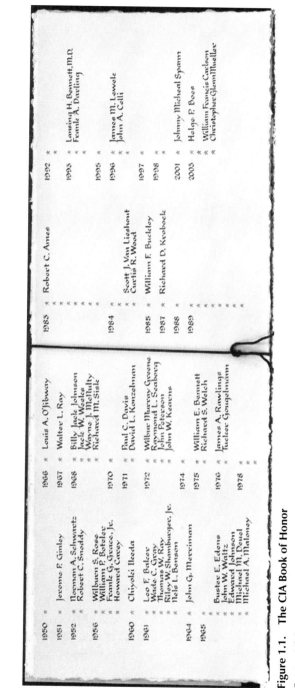

Figure 1.1. The CIA Book of Honor

Courtesy US Government

But at this juncture it should be noted that such conclusions are not absolute, and there will always be some level of probability or uncertainty involved with presenting intelligence findings (i.e., proximate reality). Nevertheless, uncertainty can be reduced and conclusion limits further defined so decision makers understand the boundaries. This must be contrasted with making decisions based on "a hunch," "instinct," "luck," "gut feel," "belief," "faith," "trust," or "hope."

Having said that, the word *intelligence* conjures up assorted notions of spying and espionage, secrets, and the world of exotic gadgetry. Yet to others, the word *intelligence* is closely associated with the Orwellian concept of "Big Brother"—a world of hardball politics and an uncompromising quest for influence.

To some degree, intelligence work is associated with these concepts, but here the study of intelligence is approached from the focus of the analytic methods that turn information into intelligence. This process is based on methods used in applied research rather than the James Bond–like devices used by cinema heroes or in the authoritarian oppression exercised by some of the world's brutal police states.

In the post–September 11, 2001 world, colleges and universities across the globe have responded to the need to develop intelligence courses for the new cadre of analysts needed to support national security. Much of what is taught in these courses will also be applicable to other types of intelligence: law enforcement, military, business, and private sector intelligence. The growth of these educational programs means that training aids are also needed to instruct new analysts in the scientific methods of inquiry for intelligence research.

> Information is the unrefined raw material used to produce finished, focused intelligence. Without information, intelligence could not exist.

INFORMATION VERSUS INTELLIGENCE

Trying to define information is difficult but not impossible. Information is like gravity and electricity, as it cannot be defined by tangible examples. Nevertheless, its properties can be observed and described, thus enabling improvement in the analytic methods that produce intelligence. The problem hard sciences face in trying to define gravity and electricity has never prevented engineers from designing and building applications that involve these phenomena. Therefore, a lack of a physical variable does not prevent analysts from producing intelligence from what we call *information*.

It is quite safe to say that every facet of our lives, whether central or incidental, is in some way related to information. We rely on an alarm clock to wake us in the morning, the newspaper to tell us what is happening in the world beyond the end of our street, the radio to alert us if rain is expected, an array of indicator lights and meters on our car's dashboard to tell us about the car's performance as we drive to work, traffic lights and signs to alert us to road conditions, and on we could go until the clock tells us it's time to lay our work aside and to go off to sleep.

Individuals, organizations, and indeed whole societies owe their survival to information. The concept of community is only possible because of our ability to collect, store, retrieve, and transfer information from one person or body corporate to another. The more complex our society, the more it necessitates the conversion of information into intelligence.

The late Colonel Russell J. Bowen, a U.S. army and CIA intelligence analyst, is attributed with saying, "Religion and intelligence are two sides of the same coin: both are institutionalizations of man's attempts to cope with his fear of the unknown; one in the spiritual realm, the other in the practical."

INTELLIGENCE DEFINED

There are many definitions of *intelligence*, and this appears to have given rise to some scholars asserting that there is no agreed position on what it means. This is simply not the case. Although there may be as many definitions as there are intelligence scholars, the differences amount to mere wordsmithing. This is because the various definitions in circulation have commonality that can be narrowed to four meanings.

Dictionaries use what is referred to as an "order of definitions" in cases where there are multiple definitions. They order the definitions by synchronic semantic analysis to clarify the different meanings. Taking this approach for the many uses of the term *intelligence* that appear in the subject literature, the term can be deduced to mean:

1. Actions or processes used to produce knowledge;
2. The body of knowledge thereby produced;[6]
3. Organizations that deal in knowledge (e.g., an intelligence agency); and
4. The reports and briefings produced for decision makers in the process or by such organizations.[7]

However, it is axiomatic that these four meanings take place in the context of secrecy. Otherwise, these definitions could apply to other forms of research. Moreover, in this book, intelligence as a process (i.e., definition 1 above) is categorized by the different functions it performs. *Knowledge* in the context of intelligence equates to *insight*, or viewed another way, the ability to *reduce uncertainty*. Insight (in other words, *advantage*), and therefore, certainty, offers mankind the ability to make decisions that enable civilizations to take better control over the "unknown." But it should be noted that insights are not produced through mystic rituals; insights are produced through processes based on sound quantitative and qualitative research methods that culminate in *defensible conclusions*. In this sense insights relate to *probability* and/or *prediction*. Expressed as an equation, intelligence could be shown as:

$$(secrecy\ (information\ +\ analysis\ =\ intelligence$$
$$\therefore\ insight \Rightarrow reduces\ uncertainty))$$

The elements of this equation will be discussed in more detail at the end of this chapter when we examine the intelligence theory.

INTELLIGENCE AS KNOWLEDGE

As a body of knowledge, intelligence deals with an adversary, a potential adversary, or a possible area of operation that is useful to managers in planning and carrying out their organization's mandate. Terms like *target, subject, person of interest, subject of interest* are some of the ways intelligence manifests itself as knowledge. To demonstrate, consider the following notional examples:

National Security Context. Intelligence from agents in the Caribbean alerts us to the imminent passage of legislation by Cuba which will legalize a multiparty, democratic political system.

Military Context. We have recently received intelligence indicating the French government has authorized a nuclear weapons test. This intelligence indicates it will take place at their Pacific test site at Mururoa Atoll during the week beginning July 16.

Law Enforcement Context. We have intelligence indicating Mack Da-Knife is planning to break into the Pine Point office of the Springfield Credit Union this Friday night.

Business Context. Intelligence suggests Nerro Entertainment is about to begin an advertising campaign in the Northeast this autumn, attempting to capture customers in the twenty-one-to-forty-one-year-old range.

Private Sector Context. Two Japanese whaling boats were observed yesterday leaving port and heading for the northwestern Pacific. Intelligence passed on by a crew member was that the vessels were aiming to catch two hundred whales for "scientific research."

INTELLIGENCE AS A PROCESS

The intelligence process is a series of procedures or steps, forming what has been traditionally termed the *intelligence cycle*. In recent years the term *intelligence process* has gained popularity over *intelligence cycle* as it has been recognized that it is not really a cycle *per se*, but a process. Nonetheless, this cycle, or process, is initiated by a decision maker who poses a question or requests advice. This is termed an *intelligence requirement* (in some intelligence agencies, such as the military, this is referred to as *essential elements of intelligence*—EEI). The intelligence requirements are forwarded to an intelligence agency and the cycle begins.

The intelligence process consists of seven steps (see figure 1.2), with the first five focusing on converting raw data[8] into finished, focused intelligence:

1. Direction setting (i.e., problem formulation and planning);
2. Information collection;
3. Data collation;
4. Data manipulation and processing; and
5. Data analysis.

This resulting intelligence is then treated with two further steps:

6. Report writing; and
7. Dissemination to decision makers (which would include provision for feedback).

Depending upon the initial intelligence requirements (e.g., the research objective), a single "loop" may be sufficient to complete the intelligence research project and provide the decision maker with the insight sought. However, in practice, further data may need to be collected with the cycle

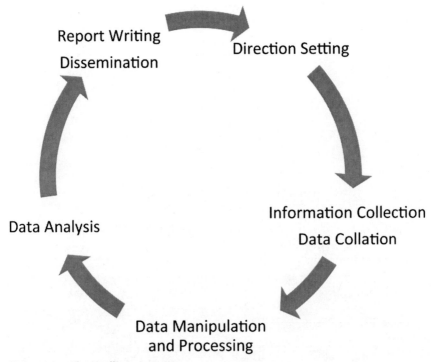

Figure 1.2.   The intelligence process.

beginning again, or the cycle may have two or more tasks being performed at once and may double back before advancing again. For instance, once the research question has been formulated and the data collection plan devised, an outline of the report may begin, and as the more readily available pieces of information flow in, a database or spreadsheet may have been constructed and the data collated.

Furthermore, even before all the data are received, some preliminary analysis may be carried out, and depending upon the results (e.g., at the collation stage which some analysts view as low grade analysis), further information may be requested (e.g., if by chance these results show the data would be inadequate to answer the research question or a serious limitation noted). This would mean that the data collection plan is revised and field operatives called on to gather more or different data, and so on.

As long as a specific intelligence operation is being conducted, the analytic process will be continuous—forming a cycle. As new information is being collected and collated, other data will be manipulated and analyzed. The resulting outcomes will be disseminated for either immediate use and/or used to set new collection goals.

The dissemination of the intelligence product can take a variety of forms. Take for instance the case of business intelligence—it could be a background history on a company or one of its executives, a diagram of a company's office layout, identification of new projects being researched, a prediction on the intended release of a new product, staff salaries, the classification and number of personnel on a company's payroll, and the like.

The intelligence cycle is not unique to intelligence research but has parallels with research cycles in other academic disciplines.[9] For instance, the research cycle that is used in applied social research shares the same pattern:

- Establish a plan for information collection and carry out initial fieldwork;
- Observe, discuss, and collect data;
- Analyze the data and write the report; and
- Distribute the report and gather feedback that can be used to formulate further dissemination strategies.

INTELLIGENCE VERSUS INVESTIGATION

In research, including intelligence research, the term *investigation* is often used. It is used to describe an inquiry into a matter under study (i.e., under investigation). Other terms are also used, for instance, an *examination*, an *assessment*, an *evaluation*, an *exploration*, an *analysis*, and so forth. However, the term *investigation* is also used in an operational sense to describe the process where officers or agents are tasked to uncover the cause of an event and the perpetrators. Examples might include an investigation into the alleged counterfeiting of trademarked fashion apparel, or an investigation into the alleged smuggling of precursory chemicals to stock a clandestine drug laboratory.

The difference between an *investigation* and *intelligence* can be summarized as follows: an investigation is usually aimed at finding the perpetrators and bringing them before a court of law to face justice, whereas intelligence is concerned with providing *insight* into the issue under study (i.e., to reduce uncertainty). Yes, intelligence can assist an investigation, as is the case of tactical intelligence, but it is not on the intelligence that the prosecution rests, it is the evidence obtained in the course of the whole investigation. So it could be argued that investigation is evidence based, and intelligence is focused on insights.

DATA VERSUS INFORMATION

Some intelligence scholars make a distinction between the terms *data* and *information*. According to *The American Heritage Dictionary for the English Language* the word *data* means "Information, especially information organized for analysis or used for the basis for a decision."[10] So in the context of intelligence research, the terms can generally be used interchangeably, as analysts rely on *data* for analysis and decision making purposes. From this perspective, *data*[11] is therefore *information*. Trying to create a distinction between these two terms, in contrast to common usage, could be argued to be an exercise in abstractness that adds nothing to understanding, yet potentially adds to confusion.

"'We have facts,' they say. But facts are not everything—at least half the business lies in how you interpret them!"[1]

1. Fyodor Dostoevsky, *Crime and Punishment* (originally published 1866), part II, chapter IV.

INTELLIGENCE THEORY

Having looked at intelligence and contrasted it with concepts such as information and data, as well as having examined the difference between intelligence and investigation and discussed intelligence as knowledge and as a process, it is incumbent to now examine the theory that underscores intelligence.

Why should we know about the theory of intelligence if we can define it, and once defined, recognize intelligence in any of its four meanings? Because theory offers both scholars and practitioners the ability to understand how and why intelligence is what it is, and does what it does. Without a theory it is difficult to posit a view about an intelligence-related phenomenon and then test that hypothesis through empirical observations to see if the results support the hypothesis, or reject it.

Although scholars have called for a theory of intelligence for decades, unfortunately until 2009 the literature was largely devoid of such theorizing. Gill, Marrin, and Phythian published an anthology of papers in 2009 that attempted to address the "missing" intelligence theory issue.[12] Among these essays were treatments by key opinion leaders such as professors David Kahn, Michael Warner, and Jennifer Sims. Although other

> Conducting intelligence research is like shining a light into a dark place.

scholars have discussed the issue elsewhere in the subject literature, these researchers were, arguably, at the time of this writing in the forefront of the debate. However, surveying the theories they advanced, it is evident that there was little consensus between the models. Nonetheless, these scholars are to be commended for advancing the debate by contributing to the discourse.

In an attempt to take their work further, this chapter proposes another theory; one that is not military- or national security-centric. This is because the world of intelligence is no longer able to operate in such neatly defined parameters. The post-9/11 world is vastly different from Cold War operations. The reasons why old demarcations have blurred will become more evident when our discussion turns to intelligence taxonomy, anatomy, and typology in the next chapter.

Grounded Theory of Intelligence

Using a grounded theory approach, which was used to develop a complementary theory of counterintelligence, this method was applied to observations made by surveying the subject literature, and then distilled to formulate an intelligence theory.[13] This theory has its roots in the definition that was put forward earlier. Although there have been many definitions of intelligence, with some scholars disagreeing on the semantic construction of "this-or-that" definition, it was possible to extract the core meaning which resulted in four unencumbered definitions presented earlier in this chapter. These definitions therefore become the first four principles on which the theory rests. Reiterating these principles, they are:

1. Actions or processes used to produce knowledge;
2. The body of knowledge thereby produced;
3. Organizations that deal in knowledge;
4. The reports and briefings produced for decision makers in the process or by such organizations; and
5. The fifth principle that is required in order for intelligence to occur in any or all of the four preceding principles is that there needs to be some context incorporating secrecy.

If the fifth principle is not present *intelligence* then becomes *research*. This is because the *knowledge* that is produced in either enterprise—intelligence and research—results in an understanding of the issue under investigation. It could be argued that knowledge leads to insight, and insight results in reducing uncertainty in decision making, but unless secrecy is involved, it is mere research. Having said that, secrecy will be context driven—what is secret for one agency may not be for another, or it may not be in a certain situation. The equation expressed earlier in this chapter presents a logical model for the theory of intelligence:

$$(secrecy\ (information\ +\ analysis\ =\ intelligence \\ \therefore\ insight \Rightarrow reduces\ uncertainty))$$

Expressed in narrative form, intelligence theory might go something like this: under the veil of secrecy, analysts obtain information and analyze it. This process results in intelligence (knowledge) and therefore provides insight to decision makers (reports and/or briefings), which in turn reduces uncertainty. This takes place within an organization (or unit within an organization, etc.) whose role is to engage in intelligence (secret research).

If the principle of secrecy is removed from the model, we can see how it transforms intelligence into research. By way of example, take the case of publicly run pharmaceutical research being conducted by a notional open laboratory for a drug to treat the effects of some common form of arthritis. Using the intelligence model just discussed, it can be seen that all the principles apply—analysts (researchers) obtain information and analyze it. This results in knowledge and therefore provides insights to decision makers (in this case, the development of a suitable drug), which in turn reduces uncertainty (that is, it gives certainty to manufacturing or other processes involved in the drug's effectiveness and/or production). But what is different is the lack of secrecy. This makes such research open and public—anyone could, potentially, access this information.

In contrast, if secrecy is applied, the research now becomes intelligence—*business intelligence* to place it in the correct typological classification (see chapter 2). Not all aspects need to be secrets, but the literature suggests at least one. Compare this supposed case of drug research to military analysts who might be researching a question about the development of a new weapons system by an unfriendly nation. In the course of their inquiries they may access open source information—say, the *curricula vitae* of certain academics in that nation who are known to be experts in the particular technology needed to develop such a weapons system. Clearly these data are freely available via the websites of the universities that employ them; but the fact that the research project is secret, the methods of analysis are classified, and other aspects of the endeavor are undisclosed, makes this *intelligence*.

Axioms

There are four supporting axioms which underpin the four principles. These can be considered to be generally accepted truths, or conditions, that allow the theory to stand. Stated, these axioms are that intelligence can be either defensive or it can be offensive, and that intelligence needs to be timely as well as defensible.

Defensive Intelligence

Defensive intelligence is concerned with providing decision makers with insights into how to deal with threats, vulnerabilities, and risks. Defensive intelligence is applicable to all five typological classifications discussed in chapter 2—national security, military, law enforcement, business, and private sector intelligence. Defensive intelligence can be concerned with several related aspects of defense—for example, these might include prevention, preparation, mitigation, damage control/response, and recovery (and perhaps other areas).

Offensive Intelligence

Intelligence can be used to assist decision makers in planning offensive missions. A simple example is that of targeting in the military. Targeting analysts use intelligence to task military assets so they can damage or destroy enemy capabilities; provide advice for immediate fire or maneuver; or support deep offensive operations. Examples of the application of offensive intelligence to other intelligence typological classifications are conceivable (see chapter 2). Offensive intelligence might include estimative or strategic intelligence, because these categories of research projects concern themselves with outmaneuvering emerging threats.

Timely

In order for intelligence to be useful, it must be provided in a timely fashion. If an intelligence report or briefing is not provided to decision makers on time, it is prima facie that the insights cannot be used. *Timely* may also include the notion of *continuous*—which is applicable in cases where an event is unfolding and updated intelligence is needed on a regular basis.

Defensible

Defensibility takes into account several related notions. These include the need for the analytic process that produces intelligence to be transparent and replicable. Transparency means transparent to those who are within the defined circle of trust, not transparent to anyone outside that circle.

The reason for transparency is to allow those reading the reports, or receiving the briefing, to reproduce the results if desired. But to do so would be most unusual; nevertheless, what is likely is that a process of review will be carried out in the same vein as an academic research report is peer-reviewed for methodological soundness. This underscores the fact that intelligence is based on the same research principles as other types of applied research—the use of the scientific method of inquiry—which is based on sound quantitative and qualitative research methods (though it is acknowledged that access to reliable information can be more difficult).

Some scholars refer to this as *auditable*. But regardless of what term is used, replication means that, say, the reader of an intelligence report can understand the collection methods, collation, and analysis techniques used (and understand why these were selected), and be able to derive similar conclusions as the analyst did from the study's findings. It is not to say that the next time the intelligence cycle is repeated the same findings will be produced—it means that if the same data and methods used to arrive at the position articulated in the report were employed, it would be reasonable to expect similar results.

Even though the environment in which intelligence operates is dynamic, that does not stand in the way of the concept of replication. Applied social research is analogously the same—rarely does an issue under investigation remain static. There are numerous independent variables in any research question that can be added, removed, or changed. This applies to intelligence research too.

If transparency and the ability to replicate an intelligence study are present, then the findings are able to be "defended." Tied to this notion are the concepts of relevancy and accuracy—these are concepts often discussed in relation to intelligence research reports. The argument here is that if the axiom of defensibility is maintained, by default, these aspects are catered for, as are the concepts of validity and reliability.

Discussion

Why does it matter that we have an intelligence theory? Because theory allows us to test propositions—questions about, say, the efficacy of certain intelligence approaches, or operational methods, or procedural practices, as well as other issues facing the profession. For instance, it allows us to test the effectiveness of intelligence concerns in terms of outcomes, outputs, and processes. As Walsh put it, it allows the development of the discipline of intelligence.[14]

So what would intelligence scholars and practitioners test with such a theory? Well, prominent among the list of possible responses is the so-called phenomenon of *intelligence failures*. For example, research questions that explore the issue of how the organization of intelligence agencies (principle 3) impacts the analytic processes (principle 1) and the resulting dynamics might result in an "intelligence failure."

Using this theory, other research questions can be formulated and tested. Take for instance these indicative examples: *Does the level of secrecy affect the operational efficiency? Is operational effectiveness contingent upon the organizational structure of the intelligence agency?* Because this is a universal theory of intelligence, it allows the context to be varied so it too can be tested. For instance, *a purely defensive approach to national security issues (which would be specified) is less effective than one that incorporates offensive measures, but in a business intelligence context, incorporating an offensive role will be counterproductive.* Using such hypotheses, scholars can then define variables and operationalize them. Take the first hypothesis above as an example: *offensive measures* could be operationalized into, say, agents, wiretaps, surveillance drones, walk-ins, or any number of other manifestations of the concept of offensive information gathering.

"I've been told all this intelligence about WMD and this is the best we've got" . . . "Don't worry, it's a slam dunk!"[2]

2. Former CIA director George Tenet's reply to then-president George W. Bush's question about the threat posed by Iraq, as cited in Bob Woodward, *Plan of Attack* (London: Simon & Schuster, 2004), 249.

As with all theories, it can then be tested empirically. Findings of empirical studies—ones based on valid and reliable data—can then guide good practice. In sum, this intelligence theory could not be described as being conceptually dense, but nevertheless it is one that articulates the five axioms that explain why intelligence practice is performed as it is, or as it should be.

It is hoped that this intelligence theory will be refined so that the theoretical base that underpins the science is better understood. "All being well, one would anticipate that, in the fullness of time, this and other yet to be articulated [intelligence] theories will spawn better policy options. These policy options will therefore be based on defensible conclusions that are grounded in empirical research."[15]

KEY WORDS AND PHRASES

The key words and phrases associated with this chapter are listed below. Demonstrate your understanding of each by writing a short definition or explanation in one or two sentences.

- Decision maker;
- Information;
- Intelligence;
- Intelligence cycle;
- Intelligence requirement; and
- Target.

STUDY QUESTIONS

1. By way of example, define the term *intelligence.*
2. Explain the difference between *intelligence* and *information.*
3. Explain the difference between *intelligence* and *investigation.*
4. List the five principles and four axioms that comprise Prunckun's intelligence theory.

LEARNING ACTIVITY

Viewing the intelligence cycle that is displayed in figure 1.2, we see that one of the key stages is *collation.* Some may argue that this stage is not only a stage in itself, but also a preliminary part of the analysis stage. Using a spreadsheet, construct a table that will allow you to collate the following data items: date, event, country of occurrence, type of military force, and impact. Now, obtain at least six newspaper articles that discuss a military event that has occurred somewhere in the world. Collate the information relating to that newspaper article into the spreadsheet, extracting a brief summary of the details (i.e., a simple qualitative description) and inserting them into the corresponding table. Having completed this activity, reflect on what you have produced and on the logic that was required to produce it. Based on this reflection, discuss why the two views of collation hold sway. That is, discuss why the collation stage can be seen as both a separate stage and part of the analytical process.

NOTES

1. George Tenet with Bill Harlow, *At the Center of the Storm: My Years at the CIA* (New York: HarperCollins, 2007), 41–42.

2. Ted Gup, *The Book of Honor: Covert Lives and Classified Deaths at the CIA* (New York: Doubleday, 2000).

3. Charles Lathrop, *The Literary Spy: The Ultimate Source for Quotations on Espionage and Intelligence* (New Haven, CT: Yale University Press, 2004), 205.

4. Ira S. Cohen, *Realpolitik: Theory and Practice* (Encino, CA: Dickenson Publishing, 1975), 41–42.

5. Mark M. Lowenthal, *Intelligence: From Secrets to Policy*, fourth edition (Washington, DC: CQ Press, 2009), 6.

6. Terry L. Schroeder, *Intelligence Specialist 3 & 2*, volume 1 (Washington, DC: Naval Education and Training Program Development Center, 1983), 2-1.

7. Christopher Andrew, Richard Aldrich, and Wesley Wark, *Secret Intelligence: A Reader* (London: Routledge, 2009), 1.

8. *Raw information* is sometimes referred to as *unassessed intelligence*.

9. Henry Prunckun, "The Intelligence Analyst as Social Scientist: A Comparison of Research Methods," *Police Studies* 19, no. 3 (1996): 70–72.

10. William Morris, ed., *The American Heritage Dictionary for the English Language* (Boston: American Heritage Publishing Co. and Houghton Mifflin Company, 1971), 336.

11. Note that the term *data* is both singular and plural.

12. Peter Gill, Stephen Marrin, and Mark Phythian, eds., *Intelligence Theory: Key Questions and Debates* (New York: Routledge, 2009).

13. In 2011–2012 I advanced a theory of counterintelligence. I saw it as a way to fill the void in the subject literature that had existed for decades. My paper on this issue, along with the method I used, was published in the *American Intelligence Journal* (19, no. 3 [2011]: 6–15), and then later published in Hank Prunckun, *Counterintelligence Theory and Practice* (Lanham, MD: Rowman & Littlefield, 2012) as well as an updated version in Henry Prunckun, "Extending the Theoretical Structure of Intelligence to Counterintelligence," *Salus Journal* (2, no. 2, June 2014: 31–49).

14. Patrick F. Walsh, *Intelligence and Intelligence Analysis* (London: Routledge, 2011), 295–97.

15. Hank Prunckun, *Counterintelligence Theory and Practice*, 48.

2

§

Intelligence Organizational Structures

This topic provides an overview of the organizational structures in which intelligence operates by examining:

1. Intelligence: a quick glance at its history;
2. Taxonomy of intelligence research;
3. Anatomy of intelligence; and
4. Typology of intelligence.

INTELLIGENCE: A QUICK GLANCE AT ITS HISTORY

The operational aspects of military intelligence are remarkably similar to those of law enforcement and business intelligence. It is not surprising then to find that these forms of intelligence find their ancestry in this genetic stock.

A cursory examination will suffice in demonstrating the lineage between these intelligence relatives. Such a comparison acts to reinforce the underlining intelligence theory—the theory developed and refined by the military and adopted by its "offspring."

The history of military intelligence dates back many centuries, and isolated examples of the craft can be cited in events from biblical times and earlier. However, it was not until the last 150 years or so that military intelligence came of age. Its official birth was registered with the formal creation of intelligence divisions within various countries' war departments.[1]

Between 1600 and until just after the Second World War, European nations began developing extensive intelligence systems but, compared to today, without great success. This is evidenced by France's miscalculation of the size of the German army at just half of what it really was at the outbreak of the First World War.

Between the First World War and the Second World War, military intelligence services expanded greatly in scope and sophistication. The only exception was that of America, whose intelligence system was to some extent disassembled. In 1929, the then U.S. secretary of state, Henry L. Stimson, advanced the now infamous dictum that "gentlemen do not read each other's mail."[2]

Stimson's comments were in response to having learned of the existence of Herbert O. Yardley's "Black Chamber."[3] Stimson is reported to have rejected any argument that justified covert code-breaking operations. He strongly disapproved of Yardley's secret activities, regarding it a low, dirty business that violated the principle of mutual trust upon which, in Stimson's view, foreign policy should be based. Stimson then shut down Yardley's code-breaking operations. History has shown the fate America suffered in the years leading up to the Second World War because of Stimson's decision to restrict intelligence to decision makers.

World War II dramatically changed any misconceptions political leaders had about the role intelligence could play; its importance to military planning and operations today is unquestioned. The intelligence offspring of military intelligence—that is, law enforcement intelligence, business intelligence, and private sector intelligence—bear the hallmarks of their parent. The two most prominent and integrally related features are:

1. Decision makers should not base their decisions on information, but rather on intelligence; and
2. Intelligence strives to answer the most pressing questions on a decision maker's mind.

Like military intelligence, national security intelligence is seen as the key to decision making. As Ransom put it: "Nothing is more crucial in the making of national decisions than the relationship between intelligence and policy, or, in a broader sense, between intelligence and action."[4]

National security intelligence is a function that is carried out by a country's foreign diplomatic service or sometimes its closely affiliated services. It also includes specialized agencies such as America's Central Intelligence Agency (CIA), Australia's Secret Intelligence Service (ASIS), Canada's Canadian Security Intelligence Service (CSIS), New Zealand's

Figure 2.1. U.S. cryptographer, Herbert O. Yardley
Courtesy US National Security Agency

Security Intelligence Service (NZSIS or SIS), and Britain's Secret Intelligence Service (MI6). These organizations centrally coordinate analysis through supporting arrangements with other agencies that process information collected from all sources—open, official, and covert.

National security intelligence could be considered a descendant of its military parent, although the link between military and national security intelligence is at times so intimate that a clear demarcation cannot be realistically declared. Many times operations that involve military intervention are preceded by covert operations involving that nation's overseas secret intelligence service.[5] It could be argued, therefore, that national security intelligence is not a descendant at all but a discipline that developed at the same time and in sympathy with military intelligence.

The responsibility of national security intelligence is to advise political leaders (and usually in the first instance, the president or prime minister) on the formulation of policies relating to a wide range of foreign policy and international political issues, including the discharging of responsibilities under a number of international pacts and treaties. To be sound and constructive, foreign policy (and where this realm overlaps with military strategy) must be based upon fact and realism. Many of the facts needed to support a nation's foreign policy are therefore provided by these intelligence agencies.

TAXONOMY OF INTELLIGENCE RESEARCH

Intelligence can be classified into four categories: basic, tactical, operational, and strategic. However, the term *basic intelligence* is a bit of a misnomer. The term infers that somehow it is elementary or simple; but it is neither of these things. Basic intelligence is concerned with analyzing historical topics. The purpose is to provide information that can be used for a variety of research projects as well as operational reasons. A simple example of the latter use is where an operative or agent is developing a "cover" or "legend" and needs factual information about what a place looked like at a particular period in time.

Some scholars have asserted that research needs to be predictive in order to be considered intelligence. However, when regarding basic intelligence, one can conclude that this is not the case. As pointed out in the previous chapter, the key aspect that differentiates research from intelligence is not prediction, but *secrecy*.

The central tenet of basic intelligence is it must be easily accessible. So, in this sense, better descriptions for basic intelligence might include *historical*, *universal*, or *collective* intelligence.[6] In chapter 11 an example of how basic intelligence can be used to produce an analysis of timelines and key dates is provided.

Tactical intelligence provides support to an operation that is either under way or about to begin. Allied to the concept of tactical intelligence is what could be considered the category of *current intelligence*. Current intelligence includes reports and briefings that keep intelligence consumers apprised of developments regarding various issues under consideration. Facts and figures produced in providing current intelligence reports could find their way into an agency's basic intelligence holdings (see discussion above).

Operational intelligence is information that contributes directly to the achievement of an immediate goal, whereas strategic intelligence relates to long-term forecasts or broader conclusions on larger objectives. In British Commonwealth countries strategic intelligence reports are termed *assessments* and in the United States they are called *estimates* (hence, the term *estimative intelligence* is sometimes used synonymously with *strategic intelligence*).

It should also be noted that a special category of strategic intelligence is called *warning intelligence*. Warning intelligence has a narrower focus, as it is usually concerned with providing cautionary advice about an event or situation that is likely to occur at a particular time or within a specified time frame.[7] These time frames are forward looking, but usually short term. Depending on the issue and the context, this time frame could be hours, days, or weeks. Perhaps months, but intuitively, if an estimate is focused on a time frame of many months it is likely to be seen as strategic intelligence. Although there is no hard-and-fast rule to what is or is not warning intelligence, judgment needs to be exercised as to what category applies. Certainly, if the issue under study is defined in terms of a year or more, it is unlikely to be warning intelligence.

The scope of these taxonomical categories is described below in point form as well as shown diagrammatically in figure 2.2. However, for the majority of purposes discussed in this book, tactical, operational, and strategic intelligence will be the focus.

Basic Intelligence

- Provides an encyclopedia-like compilation of facts and figures;
- Covers a variety of topics, issues, events, situations, places, and people spanning many decades or even centuries;

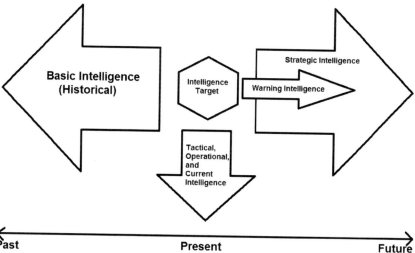

Figure 2.2.   Taxonomical categories of intelligence.

- Can be used by research analysts as well as operational personnel; and
- Can be easily accessed for quick reference.

Tactical Intelligence

- Provides immediate insight that supports a specific operation;
- Oriented toward an individual target or an activity over the short term; and
- Provides day-to-day updates on unfolding events or developing situations (i.e., *current intelligence*).

Operational Intelligence

- Is short-range or time limited but usually covers a slightly longer time frame than tactical; and
- Consists of patterns or operational mode activities.

Strategic Intelligence

- Considered to be a higher form of intelligence;
- Provides a comprehensive view of a target or an activity;
- Comments on future possibilities or identifies potential issues;
- Provides advice on threats, risks, and vulnerabilities;
- Provides warning about the likelihood of certain events or situations (see the discussion about warning intelligence above);
- Provides options for planning and policy development;
- Assists in allocating resources; and
- Requires extensive knowledge of the target or the area of activity.

Although there appears to be an obvious separation between these classes of intelligence, in certain situations a given piece of information may be relevant to more than one—say, to meet a tactical objective as well as a strategic goal.

In military intelligence, there are other categories of tactical, operational, and strategic intelligence specific to a branch of an armed service, such as combat intelligence to the army:

- Provides military commanders with advice on the threat posed by an enemy through a process known as *intelligence preparation of the battlefield* (IPB);
- Provides knowledge of an enemy's *order of battle*—that is, a list of military units, the type of equipment it carries, and the capabilities of

that equipment, as well as the location of the units and other information specific to the battlefield environment;

- Provides analysis of the weather and geographical features likely to be encountered by a commander when conducting combat operations; and
- Assists commanders in executing existing plans that are based on sound decisions—these decisions take into account the enemy's intentions, capabilities, vulnerabilities, and, therefore, the likely courses of action.

Naval intelligence has categories that are specific to its mission, such as intelligence for amphibious operations, intelligence for antisubmarine warfare, and intelligence for air operations. The air force has categories within this taxonomy of tactical, operational, and strategic intelligence applicable to its areas of concern and operations in air and space as well as information warfare in cyberspace—for instance, indications and warning intelligence, and target intelligence (i.e., target development and battle damage assessment).

ANATOMY OF INTELLIGENCE

Just as the human anatomy is comprised of different components, intelligence is also comprised of components: applied intelligence research, counterintelligence, espionage, counterespionage, and covert operations.

Applied Intelligence Research

Basic research, or theoretical research as it is sometimes known in other academic disciplines, is concerned with research for its own sake—that is, when undertaken, it has no practical application in mind. It is knowledge for knowledge's sake. The findings of such research are sometimes used later in an applied setting, but at the time of conducting the research, this was not the aim.

In contrast, applied research has a practical purpose—to offer a basis for making a decision (i.e., to provide insight or reduce uncertainty). Intelligence is in this sense applied research—it is the outcome of processing raw information that has been collected from a variety of sources—open, semi-open, official, clandestine, or covert.[8]

Once the information is in the hands of an intelligence analyst, it is evaluated and any irrelevant information discarded. The pieces of information pertinent to the matter under investigation are then analyzed, interpreted, and formed into a finished *product*. This product can take the

"Information gathered by intelligence services or compiled by the analyst is of little use unless it is got into the hands of the 'consumers,' the policymakers."[1]

1. Allen W. Dulles, *The Craft of Intelligence* (New Delhi: Manas Publications, 2007), 149.

form of an oral briefing, a written briefing, a target profile, a tactical assessment, a strategic estimate, or any number of other forms of reports. These products are then disseminated to the end user (known as the *customer* or *consumer*). The intelligence process can be summarized as analysis that leads to the production of deep, thorough, or meaningful understanding about a particular matter. An overview of some of the key intelligence topics and their potential taxonomical uses is given in table 2.1.

Counterintelligence

Counterintelligence is concerned with deterrence and detection. It is a security-focused function, but it is not security. However, security is used defensively within counterintelligence. That is, the thrust of counterintelligence is to protect an agency (or its client) from infiltration by an adversary, to protect against inadvertent leakage of confidential information, and to make secure its installations and material against espionage, subversion, sabotage, terrorism, and other forms of politically motivated violence, and the transfer of key technologies and/or equipment. It is an active model that calls on defensive as well as offensive methods of security and uses research and analysis.

Even though there is a clear distinction between intelligence and counterintelligence, this demarcation line can be thin. That is, information discovered via the counterintelligence function concerning an adversary's attempts to penetrate one's own or a partner agency can feed the intelligence side, revealing an opponent's information voids as well as highlighting their capabilities and possible intentions. So counterintelligence can be seen as both an activity that is carried out and a product that is produced to inform decision makers.[9]

Espionage

This is the classic form of information gathering dating back centuries, and it forms part of the second step of the intelligence cycle. Espionage,

Table 2.1   Selected Intelligence Topics for "Products"

Topic	Potential Taxonomical Uses
Agriculture	Strategic
Aquaculture	Strategic
Arts, Culture, and Literature	Basic Intelligence
Biographic	Basic Intelligence
Building/Construction Industry	Strategic
Civil Infrastructure	Operational and Strategic
Crime and Justice (including Civil Justice)	Operational and Strategic
Diplomatic	Strategic
Economic/Financial	Strategic
Educational	Strategic
Enemy Attacks (in various manifestations)	Warning (Strategic)
Energy	Strategic
Environmental	Tactical, Operational, and Strategic
Foreign Trade	Strategic
Geological	Basic Intelligence
Government	Strategic
Health	Strategic
Historic	Basic Intelligence
Law Enforcement	Operational and Strategic
Legal/Legislative	Strategic
Manufacturing/Industry	Strategic
Media	Operational and Strategic
Military/Defense	Tactical, Operational, and Strategic
Mining/Minerals	Strategic
Organizational	Operational and Strategic
Political (includes many subtopics)	Operational and Strategic
Religious	Operational and Strategic
Science	Strategic
Social (includes many subtopics)	Operational and Strategic
Space	Strategic
Sport	Basic Intelligence
Technological	Tactical, Operational, and Strategic
Telecommunications	Tactical, Operational, and Strategic
Transportation—Passenger and Cargo	Tactical, Operational, and Strategic
Vital Statistics	Basic Intelligence

or colloquially, spying, traditionally utilizes undercover agents. Having said that, a distinction must be made that an *agent* is someone who acts on behalf of another person or organization (e.g., a private investigator is an agent for the client who hired them to make inquiries), whereas an *officer* (or an *operative*) is someone who is charged with an authority that requires them to discharge a statutory responsibility (e.g., Boston police officer). So an agent is someone whom an intelligence officer recruits to obtain secrets on behalf of the operative's agency.[10] In such situations, the recruiting officer was traditionally known as a *case officer*, but is now termed an *operations officer*.[11]

Espionage is not a James Bond–like game. Rather, it is a serious business that can have deadly consequences. The Wall of Honor at CIA headquarters holds testimony to this fact. In 2002, it was reported to hold seventy-nine stars, each representing an officer of the agency who gave his or her life in the service of country. Forty-eight of those officers have their names listed in the Book of Honor,[2] but the remaining (at that time) were anonymous, as their services to their nation were still classified.[3]

2. Ted Gup, *The Book of Honor: Covert Lives and Classified Deaths at the CIA* (New York: Doubleday, 2000).
3. T. J. Waters, *Class 11: Inside the CIA's First Post-9/11 Spy Class* (New York: Dutton, 2006), 4.

These agents (undercover) are placed in, or recruited while in, positions that allow them to view, overhear, or otherwise obtain information that could not be gained in any other way. Generally speaking, agents betray their country because of ideological or monetary reasons; or for revenge, lust for power, or for the thrill involved; or it could be for the risk or mystique involved in the activity, or to fulfill some fantasy.[12] However, with technological improvements, more technical means of espionage (i.e., technical surveillance and unobtrusive methods) are favored over the classic use of espionage agents. It could be said that this worked reasonably well during the Cold War when intelligence agencies faced actors that were states.

On the one hand, an agent can provide an agency with a stream of intelligence for years. On the other, a defector (a specific type of spy) can only provide the agency with intelligence which is current at the time of his or her defection, though much can be learned about past operations and methods from debriefing them. Both types of spies are needed in the "great game,"[13] but each has its strengths and limitations. Although spies are often only viewed as being active in the world of national security intelligence, spies and defectors can perform these same roles in other types of intelligence work, including business intelligence and law enforcement intelligence.[14]

But since the al-Qaeda terrorist attacks of September 11, 2001, intelligence agencies recognized the importance of having agents in place to gather data.[15] One reason for the shift to technically gathered data was the comparatively high cost of running field agents and improving the reliability of the data collected (e.g., aerial and satellite photographs are not susceptible to exaggerating the truth as an agent might be. These data simply show what is there and what is not).

The events of September 11 and the subsequent terrorist attacks in Bali, Madrid, and London show that the advantages of technically gathered data were of little value against terrorist cells operating in a vastly different fashion from that of, say, a foreign government's military. This type of confrontation, and other nontraditional challenges to a state-centric paradigm, no longer applies. Nations now face threats from weak and corrupt governments, rogue states, sub-state and trans-state actors, as well as international, organized criminal groups, radical ethnic and religious groups, and right-wing political groups. All of these threats pose special data collection problems that defy a purely technical approach.

When it comes to describing *cover*—a plausible story about all facets of the operative's life—there are essentially two types: official cover and nonofficial cover (NOC, pronounced *knock*).[16] NOC is also referred to as commercial cover, when the operative works for a phantom company created and maintained by an intelligence agency. The former are personnel posing as government employees of some description, and the latter are those who on the surface have no connection with government.

The NOC operatives have been described as the truest practitioners of espionage, as they operate on their own at all times with no protection from their government. In the case of foreign espionage, if they are caught abroad, they may be tortured during interrogation and perhaps executed. If this happens, no media conference will be held, and no one will hear about the event. NOC operatives operate alone and die alone.

Nonetheless, espionage still employs audio surveillance devices, radio frequency devices, and special photographic equipment, including space-based reconnaissance satellites. The use of such devices can provide the intelligence analyst with an exponential gain in both the quantity and, under the right circumstances, the quality of the information gathered. However, it is at the peril of the intelligence agency that it neglects data collection by human sources. Because the espionage function features so heavily in intelligence work, chapter 9 explores various forms of covert and clandestine sources of information gathering.

Counterespionage

Counterespionage is concerned with detection, deception,[17] and neutralizing the effectiveness of an adversary's intelligence activities. On the surface, counterespionage presents as a form of spying—collecting classified information through a network of agents. And it is, but the difference is that it is the acquisition of data not from another nation's government or military, but from the opposition's intelligence service. It is in some ways related to counterintelligence, but differs in others. Counterintelligence could be seen as the defensive side of the craft, whereas counterespionage is the offensive side. An agency cannot have the latter without the former, so the two work in tandem.

"Counterespionage is often touted as the aristocratic sector of secret operations. In the romantic image the counterespionage man is pitted against his fellow professionals on the other side who are trying to get his nation's secrets. His job is to foil them. It is a true adversary relationship unlike the espionage situation, in which two men work together to purloin secrets. Most spy stories are not about spying but about counterspying."[4]

4. Harry Rositzke, *CIA's Secret Operations: Espionage, Counterespionage, and Covert Action* (New York: Reader's Digest Press, 1977), 119.

Counterespionage is a precise function that is, arguably, the most subtle and sophisticated of all intelligence functions. It calls for the engineering of complex strategies that deliberately puts one's agent(s) in contact with an opposition's intelligence personnel. This is done so that information can be obtained, or the adversary can be fed disinformation which will hopefully lead to confusion, thus disrupting the adversary's operations, thus allowing the perpetrator to "prosper"—and prosper can be interpreted in many different ways. False information can also be planted, so, like a "barium meal," the route that this information travels within the opposition's intelligence apparatus can be traced in order to confirm penetration by a mole or expose other security leaks or discovery information hitherto unknown. It could be argued that counterespionage could not carry out its mission without the support of the methods and practices of counterintelligence.[18]

Covert Operations

Although scholars sometimes argue that covert operations fall into a somewhat gray area of intelligence work—being neither intelligence work nor military operations—it's posited here that covert ops are in fact one of the prime policy options for which intelligence is produced. Intelligence work and covert operations should be seen as going hand-in-glove.

Intelligence professionals who view covert action as a repulsive cousin of intelligence work are misguided in their thinking. If they believe that they are only information gatherers or analysts of data, and that the outcome is simply to produce reports, assessments, and estimates, then they are naïve. They need to recall that the outcome of their activities is not related to some polite discussion about the academic merits of how the principal adversary or opposition might be influenced. It should be clear to anyone working in intelligence that the end state is to defeat the opposition and/or prevent the opposition from defeating you.

On a policy scale that ranges from doing nothing up to engaging in all-out military engagement, covert action lies somewhere in the middle. So how could intelligence professionals not see that their work might affect policy outcomes that could maintain the *status quo* at one extreme, but could escalate to inflicting death on large numbers of people or destroying entire cities? How could they not see that covert action as just another policy option, albeit a *secret policy option*, within this policy continuum? Choosing one option over another is, after all, based on the results of their intelligence work. (For more on this, see chapters 20 and 22.)

Sometimes referred to as *covert action*,[19] *special activities*,[20] or *special ops*,[21] it uses various methods of information gathering including that of research and analysis, but incorporates advice and counsel, financial and material support, as well as technical assistance to individuals, groups, or businesses which are opposed to, or working in competition with, a target or adversary.

Covert operations, or *black ops* as they are sometimes referred to, is a function by which the perpetrator uses the information it collects through espionage and observation and then analyzes it to strengthen its allies and to weaken, destabilize, or destroy its opponents.[22] A few of the tactics used in covert operations include political agitation, propaganda, election rigging, bribing high officials, blackmailing key political figures, demolishing key facilities, targeted killings, promoting rebellions and insurrections, as well as a wide variety of "monkey wrenching" tactics[23] in the physical world and in cyberspace. Listing these tactics in such a blunt way should not be interpreted as condemnation or admonishment—in the world of *realpolitik* their use is a valid policy option. But the effectiveness of covert operations is contingent upon the perpetrator's involvement remaining hidden, or at least, deniable, and this relies on intelligence work (specially, counterintelligence).

On the one hand, if a "plausible denial"[24] can be maintained, then the rewards of such ventures can be enormous. On the other hand, if the perpetrator's involvement is discovered, the consequences of this activity can be catastrophic. For example, in 1985 the French government was concerned about protests by Greenpeace regarding nuclear testing on the

> If intelligence is about secretly trying to understand or predict an event using secret methods of inquiry, then covert action is about making an impact on an event using secret tactics.

Pacific atoll of Mururoa. So on July 10, 1985, French intelligence operatives from the *Direction Générale de la Sécurité Extérieure* (or in English, the Directorate-General for External Security) mined Greenpeace's *Rainbow Warrior* in Auckland Harbour, New Zealand, with an explosive charge. The vessel sank, and the explosion killed one person on-board.

New Zealand police mounted an investigation into the incident and two French operatives were arrested, tried, and found guilty. They were then sent to prison. The other French operatives involved in the black op managed to evade capture and escaped. The incident was not only an embarrassment to the French government but carried political ramifications that affected the French government for many years. Had the operation been carried out successfully—that is, had the operatives managed to escape undetected—then the results of the op would have been much different.

Although it could be thought that agents who perform covert operations do this exclusively, it should be kept in mind that any agent who is recruited as a spy can be diverted to perform covert operations. Yes, there may be agents who specialize in covert operations, but it does not necessarily follow that they are the only ones who perform this function, especially when the situation calls for speed of execution, in which case, improvisation by using an agent-in-place may be the approach selected.

Summary

One way to summarize the relationship between the different anatomical parts of intelligence is to show it in a modified Johari Window.[25] Table 2.2 displays the four "panes" in the so-called "window" metaphor. Across the top is displayed the agency's perspective on information and along the left side is the opposition's perspective. So the intersection of each perspective shows the intelligence function that is needed by the agency in relation to each information type.

Starting at the upper left quadrant, we see that where information is known to both the agency and the opposition, it can be considered to be commonly held (e.g., open source information) and not subject to any specialized intelligence function, though it is the livelihood of the applied

Figure 2.3. Rainbow Warrior II docked at Port Adelaide, South Australia.
Photograph by author

Table 2.2 The Relationship between Information and the Various Anatomical Parts of Intelligence Viewed from an Agency's Standpoint

		Agency	
		Known	*Unknown*
Opposition	*Known*	Applied Intelligence Research	Espionage
	Unknown	Counterintelligence	Counterespionage and Covert Ops

intelligence research function. But having said that, it is important to note that analysts performing the applied intelligence research function will use information from all four quadrants. For simplicity, this function is shown in this quadrant only.

In the upper right quadrant, these data are unknown to the agency but known to the opposition, so in order for the agency to obtain this information, the espionage function needs to be employed. In the lower left quadrant we have data that is known to the agency but unknown to the opposition. In these cases the counterintelligence function needs to be exercised. Finally, the lower right quadrant shows that neither the agency nor the opposition know about this information, but both seek to acquire it. This is where counterespionage and covert operations come into play.

TYPOLOGY OF INTELLIGENCE

Intelligence is structured according to type (or class), and the typology is based on the environment in which the organization operates. There are five major classes of intelligence, including national security (which includes foreign policy and international politics), military, law enforcement, business, and private. A sixth type, emergency services (e.g., fire-fighters and search-and-rescue teams), also has intelligence cells, but although they are entitled intelligence, they do not perform the same function or perform the same level of analysis as those discussed here.[26,27] Because of these limitations, this intelligence type is not considered within the pages of this book.

It is important to note these environments can overlap—for example, an investigation into the capability of a terrorist cell may be of interest to local law enforcement agencies as well as to agencies involved in national security, the military, and some private security firms. Moreover, with regard to military intelligence, it is in some cases intimately aligned with national security because it not only informs military commanders of the intent and capabilities of an adversary but also political leaders who are responsible for authorizing the use of military force and directing strategic military policy.

In addition to the overlap or close working partnership, the same methods of operation, tactics, devices, information storage systems, and methods of analysis are used by each intelligence type. This is because information holds no bounds as to its usefulness, and a particular piece of data could conceivably be the target for more than one type of intelligence user. In other words, the primary difference between the various types of intelligence lies in the end use or general thrust of the intelligence operation.

National Security Intelligence

National security intelligence is conducted by the various branches of a nation's armed forces, foreign diplomatic service, and, depending on the country, its atomic energy authority. It is sometimes referred to as *foreign policy* intelligence, depending on the context. Nations with advanced economies generally tend to have a central agency that acts to coordinate subsidiary intelligence agencies and the collection and processing of information from all sources. Other nations, in contrast, have a unified system with one supreme agency taking on all three roles—coordination, collection, and analysis.

The types of information sought by national security intelligence analysts can be anything from the current political issues facing a foreign government; the health, education, and social structures of the country; its social problems; and its legal institutions. They may include issues concerning the availability of food production and distribution, world resources (e.g., oil and potable water), international trade relationships, world migration patterns and changes in the ethnic composition of nations, as well as the state of the global monetary order. Without a doubt, they seek also information on foreign technological developments, nuclear matters, and almost anything to do with foreign weapons production, defense industries, defense installations, and military capabilities.

However, in the post-9/11 security environment, there is a clear nexus between national security intelligence and law enforcement intelligence. This is due to the threat international terrorists pose to civil society. Pre-9/11, national security analysts and law enforcement analysts did not "... fully understand the fundamental roles and limitations of their counterparts. This led to communications errors and frustrations, ultimately leading to decreased effectiveness."[28] Although this situation has changed in the intervening years, it underscores the point that there is an overlap in these security environments.[29]

Military Intelligence

Military decisions carry a heavy burden of responsibility. As such, these decisions not only impact the lives of the fighting forces but also a nation's liberty. Military intelligence concerns itself with matters key to fighting a war: "enemy strength, capabilities, and vulnerabilities as well as information on weather and terrain."[30] In addition to dealing with these fundamental concerns, intelligence produced by the military "has to be timely, accurate, adequate, and usable."[31]

Military intelligence is decision making by commanders with regard to either the operational environment, forces (whether they are hostile, friendly, or neutral), as well as the civilian population in the operational (or potential) area. A nation's military will carry out intelligence activities regardless of whether it is at war or at peace (i.e., to prevent a surprise attack or to transition to a war footing at short notice) and at the three levels—tactical, operational, and strategic.

It would be most unusual for any army, navy, or air force not to have some form of a military intelligence capability. It may take the form of a specialist unit, or it may be part of another government service. Staff can be from the military or civilians assigned to the intelligence agency because of their particular technical or analytical abilities. Military personnel who do not have such skills are often trained at special colleges, which are set up specifically for this purpose.

Law Enforcement Intelligence

Law enforcement intelligence aims to increase the accuracy of decisions made by commanders. Intelligence provides senior officers with advice needed to make sound decisions and in this regard provides a focus on those criminal activities that would generally go undetected until they evolve into a community problem. These agencies are much wider than just police and include compliance and regulatory agencies that perform law enforcement functions (which can be quite numerous), such as immigration, customs services, and prison intelligence units.

Law enforcement intelligence also encompasses agencies engaged in combating the threat from foreign and internal subversion, espionage, sabotage, and terrorism. Other law enforcement agencies protect national security and foreign policy interests by enforcing export regulations relating to prohibited dual-use items, such as certain hardware technology, software, chemicals, and nuclear material (e.g., U.S. Department of Commerce's Office of Export Enforcement).

Depending on the agency and its mission, the analysts who staff these units may be referred to by different names, such as project officers, collators, intelligence analysts, crime analysts, criminal intelligence analysts, as well as others.[32] The reason for these different names could be historical or based on industrial agreements. The titles may also reflect some level of discrimination between the levels of critical thinking that are required to perform the tasks. For instance, a collator may perform simple analytic tasks using a few rudimental research methods, whereas an analyst may be required to show high levels of analytic thinking as well as an understanding of advanced research methodologies (and/or subject expertise, etc.).

Business Intelligence

Business intelligence is concerned with the acquisition of trade-related secrets and commercial information that is held confidential from competing firms. Business intelligence is also referred as *competitor intelligence* (or *competitive intelligence*) and *corporate intelligence*.[33] Although the media has exposed much about the unethical behavior of some intelligence practitioners, it is safe to say that a good deal of information is gathered through open- and semi-open sources. The focus of this activity can be on several levels—local, regional, national, or even international. Business intelligence is not limited to the realms of companies and corporations themselves but also can include private investigation firms that specialize in this area and intelligence agencies of foreign nations. The former often comprises spies-for-hire and the latter foreign military and national security agencies that are targeting trade and economic details and secrets.

Sometimes other terms are used by businesses to "soften" what might be seen as aggressive practices associated with the term *intelligence*. *Market intelligence* (or simply, *market research*), *product intelligence* (or *product research*), or *customer/client intelligence* are all terms that can be heard in business circles, but when used, they usually mean *business intelligence*.

"The business world is an arena of competition. Like hustling baseball clubs, campaigning politicians, and battling armies, companies are in conflict with their counterparts."[5]

5. William Sammon, Mark Kurland, and Robert Spitalnic, *Business Competitor Intelligence: Methods for Collecting, Organizing, and Using Information* (New York: John Wiley and Sons, 1984), v.

Private Sector Intelligence

To describe what constitutes private sector intelligence is not straightforward—this is because it is a diverse type of intelligence. But for the purpose of this book it will be viewed as firms and private agents who offer their services in secret research for fee or reward to the public. But a simple definition might be: private sector intelligence is secret research conducted by nongovernment entities.[34]

Although the term *private* implies an individual, there is some overlap in what constitutes private sector intelligence and what may be business intelligence or even national security intelligence. The ultimate determiner is who is contracting the "spy-for-hire."

Private sector intelligence practitioners offer a range of specialist services that go beyond the bounds of the average private investigator or private detective. Often the private intelligence practitioner comes from a background in law enforcement, military, or national security intelligence

work. Their specialties may be in background investigations or surveillance. They may have extensive training in the use of state-of-the-art optical or electronic audio surveillance equipment, and they would be
familiar with the techniques of intelligence analysis. They may offer
advice on business counterintelligence and electronic audio countermeasures (debugging). They may also specialize in providing close personal
protection for important public figures, crisis and risk management, or
business continuity planning.

Private sector intelligence agencies could (arguably) include commercial organizations that maintain databases for specialized inquiry work,
for example, credit reporting. Likewise, private sector intelligence agencies might even encompass what are called policy institutes (or think
tanks) where research agencies engage in scholarly investigation for fee-
paying clients. Private sector intelligence practitioners are being viewed
by some of their government counterparts as a viable supplementary
alternative deemed necessary in cases where resources are constrained.

Although it would appear that private investigators might dominate
this field, many, and perhaps most, are not. Because intelligence is about
structured thinking—devising research questions, formulating data collection plans, collecting and collating data, and finally analyzing these
data—many of those who practice in the area of private sector intelligence
are other than private investigators. Some, for instance, come from backgrounds in policy; some are subject area experts; some might be method-
ologists; some are data analysts; and so on.

But this is not to say that all of these private sector intelligence practitioners have policy, analytical, or research backgrounds; it is likely that
a large percentage comprise private investigators. In fact, subject area
experts are now discussing how private investigators can adapt sophisticated intelligence methods for conducting inquiries in the post-9/11
security environment.[35] By way of example, in August 2013 a symposium
on the *Privatization of Intelligence* was held in Canberra—Australia's
national capital—where leading academic and intelligence practitioners
gathered to discuss the latest developments and their implications for the
future of the profession.[36]

KEY WORDS AND PHRASES

The key words and phrases associated with this chapter are listed below.
Demonstrate your understanding of each by writing a short definition or
explanation in one or two sentences.

- Adversary;
- Agent;

- Applied intelligence research;
- Business intelligence;
- Consumer (or customer);
- Counterespionage;
- Counterintelligence;
- Cover;
- Covert action;
- Espionage;
- Essential elements of intelligence;
- Military intelligence;
- National security intelligence;
- Officer;
- Operative; and
- Private sector intelligence.

STUDY QUESTIONS

1. Provide an overview of the five major classes of intelligence as well as the functions each performs within this typological framework.
2. Describe what is different between the four major categories of intelligence—basic, tactical, operational, and strategic.
3. Identify two intelligence consumers in a military setting, and describe how they might use intelligence products.
4. What type of intelligence would it be if an analyst was tasked to assist field operatives to locate the individuals responsible for a terrorist attack on an iconic landmark? Explain why.

LEARNING ACTIVITY

Research the businesses in one of the following nations that engage in private sector intelligence—Australia, Canada, New Zealand, the United Kingdom, or the United States. List the types of services these businesses provide and compare them to the types of tasks in which you know government intelligence agencies are engaged. Examining the list of private sector activities, are you able to suggest whether any of these could be contracted to a private firm by a government agency? Speculate as to what the advantages and disadvantages might be, and then conclude with your consideration as to when and why one or more might be contracted out.

NOTES

1. For a discussion of the history of intelligence, as well as a summary of the evolution of American intelligence, see chapters 1 and 2 of Allen Dulles, *The Craft of Intelligence* (New York: Harper & Row, 1963). The history of other intelligence services can be found in texts such as those by the late Richard Deacon (pseud-onym of George Donald King McCormick, a former naval intelligence officer in World War II and a historian): *A History of the British Secret Service* (London: Mul-ler, 1969); *A History of the Russian Secret Service* (London: Muller, 1972); *A History of the Chinese Secret Service* (New York: Taplinger, 1974); *The Israeli Secret Service* (London: Hamish Hamilton, 1977); *A History of the Japanese Secret Service* (London: Frederick Muller, 1982); and *The French Secret Service* (London: Grafton, 1990).

2. David Kahn, *The Codebreakers: The Story of Secret Writing* (Toronto, Canada: The Macmillan Company, 1969), 360.

3. Herbert Osborn Yardley (April 13, 1889–August 7, 1958), author of *The American Black Chamber* (Indianapolis: The Bobbs-Merrill Company, 1931).

4. Harry Howe Ransom, *The Intelligence Establishment* (Cambridge, MA: Har-vard University Press, 1971), 3.

5. Melissa Boyle Mahle, *Denial and Deception: An Insider's View of the CIA from Iran-Contra to 9/11* (New York: Nation Books, 2004).

6. In this regard, researchers with library science degrees can be most helpful in developing basic intelligence systems. Dr. Edna Reid, Federal Bureau of Investi-gation, Washington, DC, personal communication, June 7, 2011.

7. Cynthia M. Grabo, *Anticipating Surprise: Analysis for Strategic Warning* (Lan-ham, MD: University Press of America, 2004).

8. Roy Godson, *Dirty Tricks or Trump Cards: U.S. Covert Action and Counterintel-ligence* (Washington, DC: Brassey's), 303.

9. See Hank Prunckun, *Counterintelligence Theory and Practice* (Lanham, MD: Rowman & Littlefield, 2012) for a more in-depth discussion of the counterintelli-gence function. Also see Hank Prunckun, "A Grounded Theory of Counterintelli-gence," *American Intelligence Journal* 29, no. 2 (December 2011): 6–15.

10. Ellis M. Zacharias, *Secret Missions: The Story of an Intelligence Officer* (New York: G. P. Putnam's Sons, 1946).

11. Robert Baer, *See No Evil: The True Story of a Ground Soldier in the CIA's War on Terrorism* (New York: Crown Publishers, 2002), 273; Melissa Boyle Mahle, *Denial and Deception: An Insider's View of the CIA from Iran-Contra to 9/11* (New York: Nation Books, 2004), 37, 54, and 370.

12. Fredrick P. Hitz, *The Great Game: The Myth and Reality of Espionage* (New York: Alfred A. Knopf, 2004).

13. Fredrick P. Hitz, *The Great Game: The Myth and Reality of Espionage*, 3.

14. For a detailed discussion on the case-managing agents, see Jefferson Mack, *Running a Ring of Spies: Spycraft and Black Operations in the Real World of Espionage* (Boulder, CO: Paladin Press, 1996).

15. Baer, *See No Evil*; Mahle, *Denial and Deception*; and T. J. Waters, *Class 11: Inside the CIA's First Post-9/11 Spy Class* (New York: Dutton, 2006).

16. Mahle, *Denial and Deception*, 141, 149–50, 233, 370.

17. For a discussion about how pervasive deception is in this area of intelligence work, see the personal story of a family member who lived with his father who was a spy, by Scott C. Johnson, *The Wolf and the Watchman: A CIA Childhood* (New York: W.W. Norton and Company, 2013).

18. Prunckun, *Counterintelligence Theory and Practice.*

19. Godson, *Dirty Tricks or Trump Cards*, 2–3, 304.

20. William J. Daugherty, *Executive Secrets: Covert Action and the Presidency* (University of Kentucky Press, 2004), 13–15, and note 7 at 228.

21. Frederick P. Hitz, *The Great Game: The Myth and Reality of Espionage* (New York: Alfred A. Knopf, 2004), 5.

22. See, for instance, Dennis Fiery, *Out of Business: Force a Company, Business or Store to Close Its Doors . . . For Good* (Port Townsend, WA: Loompanics Unlimited, 1999).

23. As an example, recall the case of Donald H. Segretti, who, during the 1972 U.S. presidential election campaign, conducted "opposition research" to assist Richard Nixon's reelection. "His goal was to create as much bitterness and disunity within the Democratic Party as he could." Segretti was "recruited by [H. R.] Halderman's appointments secretary, Dwight Chapin, for the political game of 'dirty tricks.'" Quotes from Tony Ulasewicz with Stuart A. McKeever, *The President's Private Eye: The Journey of Detective Tony U. from NYPD to the Nixon White House* (Westport, CT: MACSAM Publishing Co., 1990), 240. Also, during the era of the Watergate Affair there were other *monkey wrenching* campaigns planned or conducted. For details see G. Gordon Liddy, *Will: The Autobiography of G. Gordon Liddy* (London: Severn House, 1981). Finally, there is a sizable body of literature outlining how to conduct such campaigns. By way of example, these are a few indicative titles: George Hayduke, *Get Even: The Complete Book of Dirty Tricks* (Boulder, CO: Paladin Press, 1980); George Hayduke, *Byte Me: Hayduke's Guide to Computer-Generated Revenge* (Boulder, CO: Paladin Press, 2000); John Jackson, *The Black Book of Revenge: The Complete Manual of Hardcore Dirty Tricks and Schemes* (El Dorado, AR: Desert Publications, 1991); and Victor Santoro, *Political Trashing* (Port Townsend, WA: Loompanics Unlimited, 1987).

24. Peter Grabosky and Michael Stohl, *Crime and Terrorism* (London: Sage Publications, 2010), 53 and 64. See also, Richard A. Best Jr. and Andrew Feicket, *CRS Report for Congress: Special Operations Forces (SOF) and CIA Paramilitary Operations: Issues for Congress* (Washington, DC: Congressional Research Service, Library of Congress, December 6, 2006), 5.

25. The framework shown in table 2.2 was adapted from the concept developed by Joseph Luft and Henry (Harry) Ingham for understanding interactions: "The Johari Window: A Graphic Model of Interpersonal Awareness," in *Proceedings of the Western Training Laboratory in Group Development* (Los Angeles: University of California, Los Angeles, Extension Office, 1955).

26. Emergency Management Australia, *Operations Centre Management*, second edition (Canberra, Australia: Emergency Management Australia, 2001), 5–10.

27. Emergency Management Australia, *Land Search Operations*, second edition (Canberra, Australia: Emergency Management Australia, 1997).

28. David L. Carter, "Law Enforcement Intelligence and National Security Intelligence: Exploring the Differences," *International Association of Law Enforcement Intelligence Analysts Journal* 21, no. 1 (November 2012): 1.

29. Patrick F. Walsh, *Intelligence and Intelligence Analysis* (London: Routledge, 2011).

30. Joseph A. McChristian, *The Role of Military Intelligence: 1965–1967* (Washington, DC: GPO, 1974), 3.

31. McChristian, *The Role of Military Intelligence*, 3.

32. Patrick F. Walsh, *Intelligence and Intelligence Analysis*, 245.

33. Leonard M. Fuld, *Competitor Intelligence: How to Get It—How to Use It* (New York: John Wiley and Sons, 1985); and Richard Eells and Peter Nehemkis, *Corporate Intelligence and Espionage: A Blueprint of Corporate Decision Making* (New York: Macmillan Publishing Company, 1984), 78.

34. Richard Eells and Peter Nehemkis, *Corporate Intelligence and Espionage*, 185.

35. See Hank Prunckun, ed., *Intelligence and Private Investigation: Developing Sophisticated Methods for Conducting Inquiries* (Springfield, IL: Charles C Thomas, 2013).

36. A selection of the unclassified papers that were presented at this symposium was later published in a special issue of *Salus Journal* (1, no. 2, November 2013).

3

§

The Intelligence
Research Process

This topic considers the process of intelligence research by examining:

1. Problem formulation;
2. Literature review;
3. Methodology;
4. Intelligence collection plan;
5. Data collection;
6. Data evaluation;
7. Quality control;
8. Purging files;
9. Data collation;
10. Intelligence systems;
11. Data analysis;
12. Inference development and drawing conclusions; and
13. Report dissemination.

PROBLEM FORMULATION

Problem formulation is the center of intelligence research. Aristotle is attributed as saying "Well begun is half done," and this proverb resonates in the intelligence context.

But how do decision makers formulate their questions, and how do analysts arrive at the hypotheses that form the basis of their research projects? Because intelligence research is applied, the questions under investigation will have real-world origins. Whether the origin is geopolitical (national security intelligence), financial markets (business intelligence), criminal activity (law enforcement intelligence), or issues involving an adversary's order of battle (military intelligence), the questions decision makers pose are concerned with how to address the problem. Intelligence provides insight to guide possible options based on defensible conclusions derived from evidence-centered research.

Intelligence has to "educate its customers [but] this is a formidable task. . . . They have to be convinced of what it can and what it cannot achieve" by asking the right questions, at the right time and without flooding the system.[1]

8. Walter Laqueur, "Spying and Democracy: The Future of Intelligence," *Current* (March/April 1986).

Having said that, there are times when analysts are asked to provide decision makers with possible scenarios of what the future holds for a particular environment. In such cases, analysts are free to establish their own theories and create their own research questions. To do this, analysts use techniques like morphological analysis, which can generate large numbers of possible futures. Chapter 11 discusses how to perform morphological analysis.

While analysts can provide intelligence, the competing demands of tactical field commanders (e.g., executive directors in a business setting) and strategic decision makers can flood the intelligence system, thereby rendering it ineffective. Therefore, what is asked of an intelligence unit should be minimal and specific.

In the social and behavioral sciences, the term *hypothesis* is used, but sometimes in intelligence research, the term *explanation* is applied. In this book, the term *hypothesis* will be used.

For instance, a single but nonspecific intelligence request can be quite counterproductive. Take the tactical intelligence example of a field commander who asks to "see all the aerial photos relating to the terrain north of the Orrenabad Desert." This may result in boxes of assorted classified imagery being delivered hours or days after the time it is needed, rendering the information useless to the commander if he or she had intended to use it to plan an attack. Therefore, as specific requests assist intelligence analysts, they can reciprocate by yielding answers more useful to the decision makers when choosing the most appropriate operational options. Requests need to be specified using questions such as:

- How many troops does the enemy have positioned north of the Orrenabad Desert?;
- What is the enemy's order of battle?;
- Will the weather be favorable to launch a frontal assault on these positions over the next twenty-four hours?;
- If an attack is launched, can resupply of friendly forces be assured?; and
- Can air support provide both suppressing fire in the advance and evacuation of the wounded?

LITERATURE REVIEW

The purpose of the literature review is to seek research related to the issue under investigation in order to establish a conceptual as well as a theoretical context for the research project. The literature review is particular to strategic intelligence projects, though some form of abbreviated literature review may be applicable to tactical projects. Although it may be not referred to as a literature review in the tactical setting, it might be called "background," "context," or "situation." This section of the report may only be a paragraph or two as opposed to the much longer length in a strategic intelligence report. Regardless, the literature review places the analyst's project in the context of the wider issue. "No man is an island unto himself," wrote the poet John Donne, and his message, though intended for a different audience, applies to intelligence problems.[1] No research question exists in isolation from other issues. The literature review allows analysts to develop and ground their arguments, or simply to "tell their stories."

For analysts undertaking a new research project, a literature review is one of the first steps. Analysts need to discuss the theoretical base they intend to use to test their hypothesis. For law enforcement analysts, this might include deterrence theory, target hardening, rationale choice theory, differential association, social disorganization, or any number of other criminological theories.

For national security intelligence analysts, they might consider the use of sociological (e.g., Edwin Sutherland's theory on white collar crime[2]), anthropological, psychological, political science, historical, economic (especially as they relate to illicit drug importation), business (in relation to organized crime and antiterrorism), or even military theories (e.g., counterterrorism and counterinsurgency) and apply them to the issue under investigation. The theory component of the literature review should have its own subheading (or some such literary device) so the reader does not have to "hunt" for this information. Obviously, in a tactical intelligence report, theory may not feature at all, as the purpose of the report is to address a specific, short-term issue.

The literature review should talk about why the issue being studied is a problem and how the theory (i.e., the presumed relationship between the variables), if applied to the problem, could help, improve, solve, make more efficient, and so forth. Effectively, it states what benefit the study's findings might have for decision makers or how the research results might "push back the frontiers of knowledge."

Analysts need to demonstrate they have an understanding of what research has already been conducted and how the new research will either add to it, address an area that has not yet been explored, or, if a different theory is being used, why changes might be expected that are yet to be realized. The key concepts and terms need to be discussed so they can be "operationalized" in the methodology.

It is unlikely the analysts will find research that is exactly the same as the inquiries they are making (unless they are conducting a reexamination of previously conducted research). From this perspective, there will be no issue so unique the analyst cannot locate some piece of related research to inform the current scholarly investigation.

As with other forms of scientific-based research, intelligence research seeks to push back the boundaries of knowledge by adding to what is known. Therefore, the literature review allows the analyst to introduce the problem under investigation expressed as a research question or hypothesis so the reader of the report can understand how it fits within the wider context, how exploration of the problem will provide understanding, and how to provide critical insight into solving it.

On a practical point, reviewing the literature allows the analyst to gain an appreciation of some of the problems encountered in previous research and, therefore, avoid repeating the same mistakes. These problems are often in methodological issues including inappropriate sample size or selection method, invalid measurement instruments, data reliability issues, or inappropriate statistical analysis (e.g., autocorrelation of time-series data is a commonly experienced problem).

The ancillary benefit of a literature review is the discovery of secondary data sources to be used in the study. For instance, an analyst may come across quotations from field notes or interviews, and these can be used as secondary sources of data. Even though the literature review sets the scene and appears in the forefront of the research report, the analyst can always refer back to the literature in the results section of the report as a means of validating certain aspects of the findings.

METHODOLOGY

Arguably, this section is the most important aspect of the research process for strategic intelligence research. However, like the literature review, the method section is mainly applicable to strategic intelligence projects. A tactical intelligence report may state the method it used, but if it does it would be abbreviated and may only be a paragraph that summarizes the study's overall approach.

If the analyst has crafted the research question well and placed it in its historical and theoretical framework, it will guide the rest of the research process. As the methodology deals with the tangible aspects of the study, the analyst will need to define the concepts being studied so they can be measured (i.e., operationalized).

The most popular research designs include evaluations (to plan intervention programs/operations), case studies (what is going on), longitudinal studies (has there been any change over time), comparisons (are A and B different), cross-sectional studies (are A and B different at this point in time), longitudinal comparisons (are A and B different over time), and experiments or quasi-experimental studies (what effect does A have on B).

The methodology requires analysts to identify the type of data they need to collect (whether these data are from primary and secondary sources or in the form of qualitative, quantitative, or both) and how these data will be collated and analyzed (e.g., statistically or content analysis) to test their hypothesis.

Analysts also need to consider related issues including sample size and control for confounding variables (i.e., the potential that what they are observing is due to something else that they are not measuring). These extraneous influences may pose limitations for their research (e.g., possible alternative explanations for the relationship between A and B) or limits inherent in the data.

INTELLIGENCE COLLECTION PLAN

Analysts use the information collection plan as a method to design and manage their data acquisition. The plan is a simple device structuring the analyst's thinking to develop a picture of the data and what is required, where it exists, and how it can be gathered. It outlines time frames for collection of each piece of information against the time frame of the research project, allows for noting when data have been collected, and allows for any items outstanding. Therefore, the intelligence collection plan can either be a simple collection blueprint or a combined collection and management strategy. These can take the form of fishbone diagrams and data collection tables.

Because the objectives of individual intelligence agencies are so diverse, there are no set formats for collection plans. Nonetheless, a good collection plan should be couched in precise terms reflecting:

- The decision makers' *intelligence requirements* (IR) or *intelligence collection requirements* (ICR);
- The resources needed to collect data;
- What priority each data item has in relation to other data and within the whole data collection scheme;
- Who is responsible for collecting each data item; and
- The progress tracking of each data item.

The plan should be flexible to allow adjustment as changes in intelligence requirements emerge or if the objective of the research project alters.

In some agencies, there are two additional methods of acquiring information—the *statement of intelligence interest* and *essential elements of information* (EEI was discussed in chapter 1). The former is a standing request to receive finished intelligence publications when released for dissemination. Usually, analysts register their interests in topics under investigation, and the reports are forwarded to them for consideration when they become available. The standing request can be viewed as an order that remains in force until canceled, and as new material is lodged, it is sent to the analyst. EEI are time urgent requests for information (e.g., ground forces engaged in combat operations require EEI).

Fishbone Diagram

A fishbone diagram (see figure 3.1) can be used to coordinate an information collection plan.[3] A fishbone analysis is usually conducted to identify and explore cause-and-effect issues but can be adapted by analysts to help manage the collection process.

The research question or intelligence target is posted at the right-hand side of the diagram (the fish's head). The major bones of the fish are constructed by listing the different agencies or sources of information, and the minor bones subtend from the major bones, listing the data items

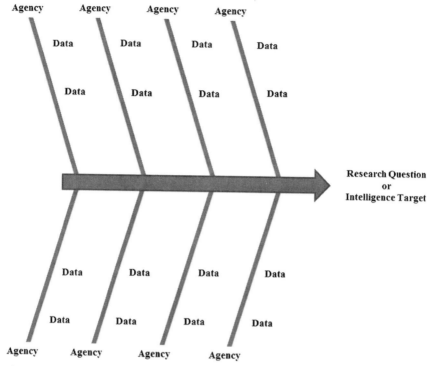

Figure 3.1. Data collection plan using a fishbone diagram.

required. As each piece of data is received, it can be crossed off the diagram with the effect of producing a visual aid as to overall progress. The pictorial information can be converted into a progress report in narrative form or as a statistical summary.

Data Collection Table

A data collection table can also be used to organize an analyst's data requirements (see table 3.1). Across the top of the table, the analyst lists the key issues for consideration. There are five shown, but other issues could be included in this row depending upon the target and the ramifications associated with the research project. Other issues could include legal constraints, administration, communication, timing, and data security.

Starting with "type of data," the analyst states the type of questions they need answered, placing each category in the row below. The analyst could substitute *type of data* with specific questions that need to be

Table 3.1.   Data Collection Plan in Table Form

Type of Data	Source	Risk	Expense	Priority
Statistical	Aerospace Facts and Figures	Nil	Cost of purchasing annual volume or online subscription	High
	Handbook of Airline Statistics	Nil	Cost of purchase	Medium
Manufacturers	World Aviation Directory	Nil	Cost of purchase	High
	Interavia ABC: World Directory of Aviation and Astronautics	Nil	Cost of purchase	Medium
Associations	Aerospace Industries Association of America	Nil	Cost of purchase	Low

answered. This is followed by completing each cell by moving from left to right across the rows. Table 3.1 is an example for the aerospace industry compiled by a notional private sector intelligence firm.

DATA COLLECTION

Information can be gathered from a variety of sources. The diversity of these sources is exemplified by category in the following lists. Many of the information sources are the same for the five different types of intelligence practitioners—national security, military, law enforcement, business, and private sector intelligence. As such, it is conceivable that a single piece of information could have application to each functional intelligence group (e.g., information relating to a terrorist cell could be of interest to law enforcement intelligence officers that have a responsibility to prevent and deter such attacks on the homeland). Likewise, the same information could be of interest to:

- National security and military intelligence if the cell is internationally based or operates from overseas;
- Business intelligence if the target of the terrorists is their industry or their facilities; or
- Groups employing private sector intelligence, such as the antinuclear lobby, to highlight the vulnerable nature of a nuclear facility in relation to terrorism.

National Security Intelligence

- Open source information (particularly the Internet);
- Clandestine operatives (official cover);
- Covert operatives (nonofficial cover);
- Recruited agents;
- Diplomatic missions and embassies;
- Surveillance planes;
- Surveillance satellites;
- Electronic intercepts;
- Defectors;
- University and independent research bodies; and
- Other government departments.

Military Intelligence

- Open source information (particularly the Internet);
- Surveillance planes;
- Surveillance satellites;
- Electronic intercepts;
- Reconnaissance teams;
- Field operatives;
- Diplomatic missions and embassies;
- Defectors;
- Prisoners;
- Civilian inhabitants;
- University and independent research bodies; and
- Other government departments.

Law Enforcement Intelligence

- The public;
- Crime investigators (detectives);
- Patrol officers;
- Police records;
- The media;
- Businesses;
- Open source information (particularly the Internet);
- Government departments and agencies;
- Informants;
- Citizens;
- Covert surveillance (physical and electronic);
- Undercover operatives; and
- Other law enforcement agencies and government departments.

Business Intelligence

- Open source information (particularly the Internet);
- A business's own internal records;
- Information supplied by other businesses;
- The media, trade, and other open source publications;
- Sales personnel;
- Customers;
- Distributors;
- Raw material and component suppliers;
- Government departments and agencies;
- A business's research and development section(s);
- University and independent research bodies;
- Market research surveys;
- Reverse engineering; and
- Covert physical surveillance (e.g., a hired private investigator).

Private Sector Intelligence

- An organization's own internal records;
- Open source information (particularly the Internet);
- Information supplied by other organizations;
- The media, trade, and other open source publications;
- Staff;
- The public;
- Government departments and agencies;
- An organization's research section;
- University and independent research bodies;
- Surveys; and
- Covert physical surveillance (e.g., a hired private investigator).

DATA EVALUATION

Evaluating information is an integral step within the analytic process and usually takes place as information is gathered. The data are evaluated according to the reliability of the source and the validity of the actual information. When evaluating information, the analyst asks many questions, including:

- How reliable is the information source?;
- Has the source provided information before?;
- How accurate is the information?; and
- How recent is the information?

With some types of intelligence, particularly national security intelligence and military intelligence, deception is a particular concern. In such cases, analysts need to evaluate data to distinguish between objective information and that tainted by bias.

> Secondary sources such as government press offices, commercial news organizations, [nongovernment organization] spokespersons, and other information providers can intentionally or unintentionally add, delete, modify, or otherwise filter the information they make available to the general public. These sources may also convey one message in English for US or international consumption and a different non-English message for local or regional consumption. It is important to know the background of open sources and the purpose of the public information in order to distinguish objective, factual information from information that lacks merit, contains bias, or is part of an effort to deceive the reader.[4]

The evaluation process firstly assesses the source's reliability and, secondly, the information's accuracy. In theory, this process is performed on each piece of information collected. However, in agencies collecting large volumes of data, this may be an automated process where a generic rating is assigned if the data are merely stored, but if used in an intelligence research project, it is reevaluated on an individual basis. Each piece of data is assigned an alphanumeric rating indicating the degree of confidence the analyst has in that piece of information. This system is universally known as the *admiralty ratings*, and a system based on this, but enhanced by the author, is shown in tables 3.2 and 3.3.

Imagine a field operative obtains a piece of information from an agent in place, but this agent is a new source that has never been exploited before. The reliability for this piece of information would therefore have to be H—the reliability cannot be judged. If the information obtained

Table 3.2. Source Reliability Codes

Code	Descriptors	Admiralty Ratings
		Estimated Truth Based on Past Reporting
A	Completely Reliable	100%
B	Usually Reliable	80%
C	Fairly Reliable	60%
D	Not Usually Reliable	40%
E	Unreliable	20%
F	Unintentionally Misleading	0%
G	Deliberately Deceptive	0%
H	Cannot Be Judged	50%

Table 3.3.   Information Accuracy Codes

	Admiralty Ratings	
Code	Descriptors	Estimated Probability of Truth
1	Confirmed	100%
2	Probably True	80%
3	Possibly True	60%
4	Doubtful	40%
5	Improbable	20%
6	Misinformation[1]	0%
7	Disinformation	0%
8	Cannot Be Judged	50%

1. Note that there is a difference between *misinformation* (which is unintentional) and *disinformation*—which is outright deception (i.e., intentional).

came from a database that has been the source of previous information (i.e., from another agent) and has proven to be truthful in almost every instance, then an accuracy code of 2 would be assigned. The combined code would be printed on the document to show its overall rating. Customarily, accuracy rating precedes the reliability code—for instance, H-2. Having said that, it needs to be pointed out that the ratings must be logical; assigning a rating of, say, E-2 (unreliable source but probably true) or G-3 (deliberately deceptive but possibly true) would raise questions in the consumer's mind about whether the evaluation process was rational.

Regarding the accuracy code 6 of table 3.3—misinformation—the analyst should be cognizant they may obtain data that are *unintentionally* incorrect, illogical, or contradicted by other sources. In these cases a code of 6 is appropriate. As for disinformation (code 7), these are data that are shown by other sources to be *deliberately* false or misleading (i.e., provided for the purposes of deception).

Although the admiralty ratings represent an objective position, they are derived through a subjective process, as judgment plays a key role. When assigning a rating, the analyst must consider such things as the accuracy of previous information provided by the source and the source's field capabilities (i.e., does the source have access and the ability to obtain what has been delivered?). Evaluation is a difficult process but an important one, as personal or agency bias can adversely affect the results of an intelligence research project. A good evaluation is the result of the source's reliability being evaluated independently of the credibility of the information. Information discovered to be irrelevant to the issue under investigation should be disposed of according to the analyst's agency document destruction policy (e.g., shredding).[5,6,7]

QUALITY CONTROL

To assist the analyst in determining the relevance of any particular piece of information to the research project, a method of labeling each data item is needed. If the analyst can correlate data with other sources of information thought to be reliable, then, theoretically, this will increase the chances the data are correct. Some intelligence agencies require that to assign a data item an A-1 rating, the data need to be verified independently against two or more sources.

However, as sound as this procedure appears, it can be subject to phantom or ghost data—data that are self-validating. By way of example, the author was once involved in an intelligence operation, and the operational team obtained document A, which cited the occurrence of a particular meeting involving one of the operation's targets. If true, this information would have led to a major breakthrough by piecing together a hitherto incomplete picture of a highly illegal industry. So the intelligence team set about checking this information with other sources and as a result they came up with two independent confirmations—documents B and C. But perceptive observation told the team something seemed odd. There were a few words in documents B and C (the additional confirmatory documents) that were similar. Checking the sources (a time-consuming task) revealed, to the team's astonishment, all three reports (A, B, and C) were based on a single report, which in turn was authored by another source! It should be noted that although these were documentary reports, the same holds for oral reports by an informant or undercover agent. In this case, the data were self-validating. If an intelligence assessment was prepared on these data, the consequences do not need to be spelled out. But given the guidelines, the assessment would have been correct. It is worth noting that data validation systems, such as the admiralty ratings, should be only guidelines, not rules.

PURGING FILES

Another aspect of quality control is purging redundant data. Intelligence systems tend to accumulate data very quickly, and in turn, these data contain material which is not always central to intelligence projects. This is mainly due to the analysts' inability to predict what data will be needed in the future; therefore, if a piece of data could be used, it tends to be retained.

However, at some stage, an assessment of the agency's holdings should be done (i.e., at the end of a project or operation). During such an audit, material lacking accuracy, relevance, timeliness, or completeness should be purged from the system. Retaining this material does nothing but detract from the quality of the overall database, and if it is used in an intelligence assessment, the data will only reduce the accuracy of, and possibly harm, the final recommendations. In some jurisdictions, having irrelevant data could be a breach of the criminal law.

When data lacking these qualities are discovered, they should be subject to upgrading. But if the cost of validating or verifying this information proves to be in excess of the potential value of the anticipated result, the data should be removed. This process should be done on a project-by-project or operation-by-operation basis. In this way, the validating and verifying is done with the most ease and effectiveness, while the project is still fresh in the minds of the analysts. Likewise, the accuracy of the index should be maintained daily—corrections and updates should be done as they are encountered. This makes for efficient and effective management of time and of the database.

A biyearly quality control exercise could be done on the whole database. The criteria cited in table 3.4 can be applied to the data and upgraded and/or discarded. In some intelligence agencies, if data cannot be judged, the information is sent to the originator of the report, who is

Table 3.4.   Some Suggestions for Retaining or Purging Information

Accuracy	How reliable is the source?	
Relevancy	Does it supply the intelligence unit with information necessary to complete its project?	
Timeliness	Does each piece of the information relate to the current project or operation?	
Completeness	Externally Obtained Data	Is the document's source stated clearly? If not, can it be determined and inserted to be made complete?
	Internally Generated Data	Is the source of the facts referred to in the document footnotes or otherwise differentiated? If not, can the reader be provided with the source or authority for these data items?
	Considerations for All Data	Are inferences and comments made by analysts clearly indicated in the document so as not to be confused with verified facts?

asked to upgrade it or otherwise make comment on it. The upgraded data are then entered into the system, replacing the old. If it cannot be upgraded, the data are destroyed.

DATA COLLATION

Collating takes place after the data have been evaluated, whereby the analyst brings together the disparate pieces of information so the data can be subject to some form of analysis. The collation process also acts to remove irrelevant, incorrect, or worthless information, which may be collected due to error, misdirection, or compulsiveness. Such data should be destroyed immediately after being identified; otherwise, it will not only cause congestion in the intelligence database but also may place the analyst and the agency in legal jeopardy if a judicial officer tasked with overseeing the agency's operations deems holding such data to be contrary to law.

The remaining data are then stored to be retrieved easily by the analyst. If the data are in an unstructured form, this can be done by:

- Registering the information;
- Indexing;
- Cross-referencing; and
- Keywording.

These data can then be filed so that an electronic database can be interrogated by the analyst using one or more analytic methods. Typically, unstructured data are stored on a mainframe or an enterprise server that is accessed by the analyst's workstation computer (which in turn is likely to be networked with other workstations and servers within the agency).

INTELLIGENCE SYSTEMS

Basic Data Storage and Retrieval Concepts

Information is stored for the purpose of retrieval and not stored for the purpose of warehousing documents. This may seem obvious, but in practice, this is not always the case. Information retrieval is the selective and systemic recall of warehoused information; hence, the information must be stored logically, or it may become lost "in the system."

Storing and retrieving data effectively does require technology. At the center of many intelligence systems is some form of index. The purpose of the index is to facilitate the retrieval process, but by its nature, an index imposes limits on how information can be retrieved. A superior indexing system allows for subjective, intuitive searching that is fundamental to intelligence research. The most common types of indexes are those listed below, and a good intelligence database should be capable of producing all.

Author Indexes. People, organizations, corporate authors, government departments and agencies, universities, research foundations, and so forth.

Alphabetical Subject Indexes. Headings, subheadings, cross-references, and qualifying terms.

Keyword in Context. A system specific to computerized systems. It operates by selecting keywords appearing in the body of the text (e.g., the keyword *blackmail* in a corruption report).

Hierarchical Indexes. Data items are arranged in a hierarchy, starting with topics of general scope and progressing to more specific topics.

Permuted Title Indexes. Systematically rotating the words in the title of a document/file. The success of the permuted index depends on the accuracy of the original author (or where no title is given, e.g., ad hoc papers and memos, the collector, writer, or researcher) creating the title to reflect the content of the document or file. This may be difficult in the cases where a document/file covers a number of topics.

Sound Indexing. *Soundex*[8] is an indexing method based on sound where names are encoded for retrieval.

Soundex-Based Systems

Soundex is a way of searching data where the exact spelling is unknown. This is particularly valuable as names can be spelled several different ways. Because a soundex-based search seeks the phonetic spelling of a word, a search will usually be more reliable if carried out for single words rather than combinations of words, such as titles and phrases (depending on the software package used).[9] The soundex method of indexing involves four steps[10]:

1. Retain the first letter of the name and drop all occurrences of a, e, h, i, o, u, w, and y in other positions.
2. Assign the following numbers to the remaining letters after the first: $1 = b, f, p, v; 2 = c, g, j, k, q, s, x, z; 3 = d, t; 4 = l; 5 = m, n; 6 = r$.
3. If two or more letters with the same code were adjacent in the original name (before step 1), omit all but the first.
4. Convert to the form "letter, digit, digit, digit" by adding trailing zeros (if there are more than three).

Internet-Based Systems

Internet-based search engines use a two-step process for indexing material posted to websites. The first step is "web crawling" or "spidering." Here a software agent methodically browses or crawls the Web and makes copies of web pages by following the hyperlinks on each page. These copies then form the basis for the next step, which is indexing.

However, because the algorithms used to construct Internet search engine indexes are commercial-in-confidence, they cannot be reviewed here. Nevertheless, it is widely understood these algorithms are based on interdisciplinary concepts borrowed from the likes of linguistics, cognitive psychology, mathematics, and informatics (and perhaps others).

As Internet-based search engines are commercial ventures and the companies running these services have to compete in a very competitive marketplace, the owners try to balance the thoroughness provided by a search request with retrieval time. In doing so, the search engine needs to give the researcher the most relevant web pages. It is in deciding what is relevant that each search engine differs in its construction of its search algorithm—hence, the interdisciplinary and multidisciplinary approach to indexing.

When it comes to images, specifically images of people, indexing takes a different turn. With the growth of facial recognition databases comes software that will allow analysts to search these holdings. These methods of searching for images of people's faces are not based on the principles of indexing, but on algorithms that compare features of a sample image with those contained in the database.

DATA ANALYSIS

Information is analyzed to draw conclusions about an activity, a person(s), a group(s), or an organization(s) at the center of inquiry in order to provide insight for the decision maker (i.e., *information* is analyzed to produce *intelligence*). The analytical process can be described in a number of steps that comprise:

- Examining the collected data;
- Sorting facts from opinions (i.e., evaluation of the information);
- Developing inferences (by means of statistical or other methods);
- Discussing the strengths and limitations of the various inferences (e.g., based on probabilities); and
- Drawing conclusions from these results and making recommendations in accordance with the decision maker's intelligence requirements.

Despite the complexity of data analysis, the process of analyzing data is generally straightforward. Intelligence research projects mostly perform analyses following a three-step process:

1. Preparing the data by "cleaning" errors and anomalies that may have crept in during the collection phase;
2. Organizing the data so it can be described statistically if quantitative and in other ways if qualitative; and
3. Testing the research hypothesis (or model) using either statistical tests or specialized analytic techniques (depending on the type of data).

The first step—preparing the data—starts with some form of logging that allows the analyst to check what has been collected against the information collection plan. Once assured all data items are present, the analyst checks the data for accuracy, which involves entering the data into a software program. Depending on the type of data being processed, it may mean "double entering" the data to ensure there are no mistakes in data entry or checking that the data item is within a specified range or in a certain format. Software programs designed for data analysis will usually have "error trapping" subroutes that highlight such errors or allow the analyst to set parameters depending upon the data being manipulated.

The purpose of using a software package is to provide some form of organization to the data so descriptive analysis can be performed. Most analytic software has this function and will produce a range of descriptive statistics forming the basis for testing the research hypothesis. Descriptive statistics are not only part of both quantitative research and qualitative projects, but also they are important as they describe what is going on. Such descriptions include graphical representations (e.g., pie diagrams, bar charts, or line graphs) coupled with a narrative discussing the various quanta.

INFERENCE DEVELOPMENT AND DRAWING CONCLUSIONS

The final phase of the intelligence research process is testing the research hypothesis. This can be done using statistics or another technique if the data are qualitative. The overall purpose is to draw conclusions allowing the analyst to extrapolate meaning from the data to some level beyond the immediate. Based on the sample data, an analyst may form some inference that could be applied to the general population from which the sample was derived. Alternatively, an analyst may use inferential statistics to form a judgment about the probability that the observations regarding two groups are the result of some variable acting on one of the groups, and the difference would not have occurred if chance was the only factor.

An *inference* is a statement (or proposition or judgment) drawn from data that have been subject to some form of analysis. In this way, the statement follows logically (either deductively or inductively) from the data. Inferences may be based on as little as one or two premises, or they can feature many premises in a cascading fashion depending on the data and the original research question. A simple example of an inference based on a deductive process consisting of two premises is:

- People are criminals because they have been found guilty of breaking the law.
- Mack DaKnife has been found guilty of breaking the law.
- Therefore, Mack DaKnife is a criminal.

You will note in the deductive reasoning process, the analyst moves from specific data items to a general position. In inductive reasoning, the opposite takes place—the analyst starts with a generalized position and moves to the specific, as seen in this example:

- Country Q is similar to Country X.
- Country Q provides a safe haven for terrorists.
- Therefore, Country X provides a safe haven for terrorists.

In the first example, the two premises are both true; therefore, the inference is valid. Although the premises in the inductive reasoning example are not false, there are many other particulars (i.e., variables) that need to be considered before the analyst can draw the inference that has been presented about Country X's providing a safe haven for terrorists.

The striking feature of the two approaches is that inferences developed by using deductive reasoning do not suffer the same degree of uncertainty as those developed by inductive reasoning. Nevertheless, inductive data analysis is useful in certain situations, such as exploratory qualitative case studies.

A general (logical) truth of a matter cannot be established by examining only one, or a few, of the variables, as there are potentially a very large number (some might say infinite) of variables that need to be weighed in an inductive argument.

Contrast this with a deductive argument where the premises and their conclusion are so integrally related that if the premises are true, then the conclusion must also be true. For inductive reasoning to be valid, it requires all the initial premises to be true, and as this is not possible, an inductive argument can only establish a degree of likelihood or probability.

> Deductive reasoning produces either a valid argument or an invalid argument. Inductive reasoning can only produce a cogent, or sound (i.e., probable), argument.

Interpreting information is a cognitive process based on general knowledge, life experience, common sense, and data collected in relation to the issues under investigation. This process involves identifying new issues and postulating the significance of these issues, which might include being viewed from the perspective of the target.[11] The "so what?" questions can be answered in the interpretative process: what does this information mean in relation to the issue under study? The answers provide a useful starting point to formulate future courses of action and make recommendations.

> It is equally important to discover evidence of disagreement in the data as it is to find evidence that supports the research hypothesis. Differences add density to the conclusions drawn. Therefore, analysts should look for outliers and rival explanations in the data.

REPORT DISSEMINATION

Dissemination is the term commonly used for the last phase of the intelligence cycle where the product (e.g., a report or briefing) is delivered to the decision maker. Because intelligence reports vary in size from one-page briefings (e.g., tactical or operational reports) to book-length studies (e.g., strategic assessments), it is hard to categorically state who the end user will be within an agency or even at what level of political leadership it may ultimately end up (recall also that some intelligence reports are only disseminated to other analysts).

In the case of business intelligence, an intelligence report may be received by a board of directors (i.e., nonexecutive level), or it might be considered by the corporate executives who make day-to-day decisions about the business. Then again, it may only be read by a division-level manager who simply needs the insights to plan production schedules.

In the case of national security intelligence, strategic reports, unless in the form of a national estimate (essentially a synopsis of a much larger document), will not be read by a decision maker. In all likelihood, strategic reports will only be read in their entirety by one of the decision maker's staff or advisers. These personnel will extract what they consider the key findings (and these may not be the key findings of the study) and brief the decision maker themselves. A strategic intelligence study may even be destined for a subject specialist, who may be a fellow intelligence analyst within another agency.

KEY WORDS AND PHRASES

The key words and phrases associated with this chapter are listed below. Demonstrate your understanding of each by writing a short definition or explanation in one or two sentences.

- Admiralty rating;
- Data analysis;
- Data collation;
- Data collection table;
- Data gathering;
- Deductive logic;
- Dissemination;
- Fishbone diagram;
- Inductive logic;
- Inferences;
- Information accuracy;
- Information evaluation process;
- Intelligence collection plan;
- Intelligence collection requirements;
- Intelligence systems;
- Methodology;
- Misinformation;
- Operationalize; and
- Source reliability.

STUDY QUESTIONS

1. Identify two reasons for conducting a literature review, and explain their importance to the overall intelligence research process.
2. What are some of the benefits of data collation? List three issues, and describe the positive impact of each.
3. Why formulate an information collection plan? Discuss what could go wrong if a plan is not incorporated into an intelligence research project.
4. Explain the purpose of using a software package for collating data.
5. Outline three methods for indexing data. Then choose one and describe its strengths and its limitations.
6. Outline some reasons why an analyst should review data held in the agency database or file system for purging.

LEARNING ACTIVITY

Review the difference between inductive and deductive reasoning. Construct two simple inductive arguments and two simple deductive arguments. Provide a short explanation as to why each is so.

NOTES

1. John Donne, Meditation XVII (1622), http://www.online-literature.com/donne/409/ (accessed November 11, 2008).

2. Edwin H. Sutherland, *White Collar Crime: The Uncut Version* (New Haven, CT: Yale University Press, 1983).

3. Also known as an "Ishikawa" diagram by the Japanese academic Kaoru Ishikawa to promote quality management process. Although this type of chart is generally used for cause-and-effect analysis in industry and commerce, it can nevertheless be adapted for other purposes, including, as shown here, information collection plans for intelligence analysts.

4. U.S. Department of the Army, *FMI 2-22.9: Open Source Intelligence* (Washington, DC: Department of the Army, 2006), 2–10.

5. International Association of Chiefs of Police, *Law Enforcement Policy on the Use of Criminal Intelligence: A Manual for Police Executives* (Gaithersburg, MD: IACP, 1985).

6. Jack Morris, *The Criminal Intelligence File: A Handbook to Guide the Storage and Use of Confidential Law Enforcement Materials* (Loomis, CA: Palmer Press, 1992).

7. Henry Prunckun, *Information Security: A Practical Handbook on Business Counterintelligence* (Springfield, IL: Charles C Thomas, 1989).

8. Soundex was originally developed by Margaret K. Odell and Robert C Russell, cf. U.S. Patents 1261167(1918) and 1435663(1922).

9. Henry Prunckun, *SpyBase* (Adelaide, Australia: Slezak Associates, 1991).

10. Donald E. Knuth, *Sorting and Searching,* volume 3 of *The Art of Computer Programming* (Reading, MA: Addison-Wesley, 1973), 391–92.

11. Michael Scheuer [pseud.], *Through Our Enemies' Eyes: Osama bin Laden, Radical Islam, and the Future of America* (Washington, DC: Brassey's, 2002).

4

🔅

The Scientific
Method of Inquiry

This topic discusses how the scientific method of inquiry is used in intelligence research by examining:

1. Scientific research methods;
2. Reasoning;
3. Probability;
4. Hypothesis testing;
5. Constructing a research hypothesis;
6. Variables;
7. Operationalizing variables; and
8. Measuring variables.

SCIENCE-BASED RESEARCH METHODS

Scientific methods of inquiry are the techniques and processes used by researchers to:

- Investigate social, criminological, psychological/behavioral, political, military, business, and economic phenomena as well as allied areas of interest;
- Acquire new knowledge in these areas; or
- Set right or incorporate previously gained knowledge.

It involves the systemic analysis of phenomena and logical problem solving. The aim is to provide insight through a transparent process in order to reduce uncertainty.

The methods are considered *scientific* because they are based on empirical evidence that can be observed (directly or indirectly) and measured. These data are subjected to the established principles of logic and must

---

Reducing uncertainty is a problem well noted by military commanders like the Prussian general Carl von Clausewitz (1780–1831), who articulated such concerns in his landmark treatise *On War*.[1]

1. Carl von Clausewitz, *On War*, J. J. Graham, trans. (New York: Alfred A. Knopf, 1993).

---

be repeatable, as they are in the physical sciences (e.g., chemistry and physics). Scientific methods of inquiry are founded on the steps of the intelligence cycle—problem formulation, data collection, data collation, analysis (including hypothesis testing), and dissemination.

Although the focus of intelligence research differs from other fields of inquiry, the features of scientific inquiry remain the same. Like other researchers, intelligence analysts pose hypotheses to explain phenomena and plan methodological approaches to study real-world problems.

One prominent feature of scientific inquiry is the objectivity of the method, so bias is reduced when collecting and interpreting the data. In academic fields such as sociology, criminology, psychology, history, anthropology, political science, military science, economics, education, as well as specialized fields like library science,[1] the findings are shared through publication in scholarly journals and professional conferences. There are a few reasons for sharing the research results, including:

- Knowledge is cumulative, and all scholar researchers benefit from publication (i.e., increasing the sum of knowledge); and
- Colleagues can critique the methods and interpretations of a study to exercise some level of quality control through peer review.

But this is not the case with intelligence research because it is secret research.[2] The audience is key decision makers responsible for protecting and preserving a way of life for many people; hence, the research is not intended for public airing, even within academic circles. For instance, the inability to critique methods and interpretations of a study is a drawback with intelligence research, but in its place, analysts could consider finding ways within their agency to have their work peer reviewed. This requires more than proofreading by a supervisor but could be done through a work-in-progress forum comprising fellow analysts where researchers present their methods and findings.

REASONING

Deductive Logic

Deductive reasoning (or deductive logic) uses arguments to move from general statements (the premise) to a specific position to draw a conclusion. The key feature of deductive reasoning is that the premises used to create the argument must be true. The premise consists of one or more propositions as well as another proposition referred to as the conclusion. Because the premise is true, the conclusion must also be true.

To assess whether a deductive argument has been constructed correctly, the analyst must ensure the argument is valid and sound. An argument is valid if the conclusion is a logical consequence of the premise. It will be sound if it is valid as well as having a true premise. Consider the following example:

- All people breathe air (the major premise).
- Fortune 500 executives are people (a minor premise—there may be others).
- Therefore, Fortune 500 executives breathe air (the conclusion).

Note, however the *validity* of a deductive conclusion is not affected by the fact that the premise is not true, as the conclusion will still be valid. Consider this example:

- Some Fortune 500 executives live in rural towns.
- Some rural towns have been abandoned and are ghost towns.
- Therefore, some Fortune 500 executives live in ghost towns.

Although this argument is *valid*, it is not *sound*, and for the argument to be sound, the conclusion must be based upon a premise that is true. But analysts will not always have both validity and soundness in their arguments, and if they did, there would be little cause to conduct research. When contrasting deductive logic with inductive logic, one finds rather than the arguments moving from general statements to a specific position to draw a conclusion, the reasoning moves from the particular to the general, and this is usually the basis of intelligence research.

Inductive Logic

Inductive reasoning makes generalizations based on individual observations (e.g., phenomenal patterns). It is used to attribute traits or relationships, or to formulate rules or theories. For example:

- Premise: There is a shortage of heroin for sale on the streets of Sydney.
- This infers the general proposition or major premise: There is a shortage of heroin throughout all of Australia.

Although inductive reasoning is valuable for posing a hypothetical position, it cannot be seen in the same light as deductive logic—both valid and sound. For the conclusion to be sound, the premise needs to be true, and it is not likely an analyst will have every piece of information to construct such an argument, so the premise is true as the number of particulars may be in the hundreds or thousands.

Belief is like *faith*; both are expressions of opinion or conviction that something is true but have not been subjected to rigorous proof based on evidence. Belief and faith are, therefore, not recognized research methodologies.

PROBABILITY

Because inductive logic can be uncertain, analysts use probability when drawing conclusions: "this is the most likely truth based on what is presently best evidence." An analyst who tests a particular theory will construct a hypothesis based on the inductively produced theory (or an inference); then using an appropriate method, the analyst will subject the research question to scientific inquiry.

Consider a study into a heroin shortage in Australia where illicit drug users in a few areas of Sydney experienced difficulty in obtaining heroin. A number of direct and indirect measures of availability were collected, and these data were subjected to analysis. The inference was that there was less heroin on the streets.

Some scholars extrapolated from these particulars an argument that the shortage was due to law enforcement efforts.[3] The conclusion was valid, but the argument failed to consider a number of other propositions making the premise not true, including a Taliban-imposed restriction in Afghanistan where opium is grown and diversion of Southeast Asian heroin (destined for the Australian black market) to Europe.

The initial conclusions were no doubt valid, but their soundness was questionable. It is likely that law enforcement played some part in the heroin shortage, but the probability that it was the sole basis for such results was very unlikely. Therefore, what analysts need is some means of assigning a level of probability to their inferences, including statistical techniques.

Tests of statistical significance can determine the likelihood that a particular result could have occurred by chance. Chance occurs when forces other than the independent variable act on the dependent variable. If it is unlikely that the result occurred by random variation, then the analyst can conclude the independent variable was responsible. The acceptable levels of probability (significance level) are 5 in 100 ($p < 0.05$) and 1 in 100 ($p < 0.01$) (this refers to the probability of rejecting the null hypothesis—see "Constructing a Research Hypothesis"). Therefore, if these results are obtained, the analyst can conclude that the study was significant at the acceptable levels, thereby supporting the hypothesis.

Another way of looking at probability is to derive a ratio between the number of times an event will occur and the number of opportunities for that event to take place. For instance, after assessing historical data in relation to terrorist attacks, an analyst calculates that an attack occurred once every three months. This can be expressed as a ratio of 1:90. In this form, probabilities can be determined—for example, a ratio of 1:10 is far more likely than a ratio of 1:100 or of 1:1,000.

While numeric (quantitative) data can be subjected to statistical tests of probability and ratios constructed, analysts should not forget that nominal and categorical data (qualitative) can be tested using the chi-square. This technique will be discussed in chapter 16. Otherwise, qualitative data can be assessed through the use of a set of scale-like terms. For instance, the U.S. intelligence community uses the terms listed in table 4.1 to reflect probability, which range from *remote* to *almost certain*. Note that the term *certain* does not appear in the scale.

Table 4.1. Categorical Terms for Attributing Probability

Remote	Very Unlikely	Unlikely	Even Chance	Probably/ Likely	Very Likely	Almost Certain

HYPOTHESIS TESTING

A hypothesis is the statement putting forward a proposition about a fact or set of facts. This statement, constructed by the intelligence analyst, is the basis of the research question or statement of guiding principle. Although a hypothesis can be created several ways, it is most likely the result of some form of inductive reasoning.

A problem is identified and a hypothesis framed, which usually starts out as the research question. The question might ask:

- How much can we expect organized crime to grow in three years' time?;
- What industries will organized crime be expected to engulf in the coming five years?; and
- Will their methods of operation differ from those being used today?[4]

These questions, as they stand, are too broad to test using the scientific method. They are tentative questions derived from the issues raised by decision makers, social observers and commentators, or the subject literature. These questions may form part of the rationale for the study, but they need to be refined to be testable and falsifiable. Take for instance the question that asks: what industries will organized crime be expected to engulf in the coming five years? Analysts might postulate that the nightclub industry may be the target of organized crime based on observations of local or interstate field operatives or the experiences discussed in the literature from overseas. Using this theory, they craft a testable hypothesis to be the subject of their inquiries. It can be in the form of a question, a statement, or an "if" statement, for example:

- Will the nightclub industry be the target of organized crime?;
- Nightclubs are an attractive target for takeover by organized crime; or
- If organized crime is looking to expand its influence, then the nightclub industry will be a target.

Any of these hypotheses are acceptable, as they allow the analyst a clear focus for study. Each needs to consist of a dependent variable and an independent variable to allow the former to be manipulated by the latter. As a hypothesis must be testable, it must also be falsifiable. Consequently, a hypothesis can never be "proven," but it can be "supported," and it is to what degree a hypothesis is supported that determines its acceptance. So statements such as "and this is strong evidence in support of the hypothesis" or, alternatively, "these data do not support the hypothesis" (i.e., do not reject the null hypothesis) are common when discussing research findings.

CONSTRUCTING A RESEARCH HYPOTHESIS

To construct a research hypothesis, analysts must state what they consider will be confirmed by their research. Think of a hypothesis as a theory, or an inference. It is the result of taking various pieces of information and arranging them so that a "picture" emerges. This picture is the hypothesis—it is an explanation of what has happened (past), what might be going on (present), or about to happen (future). Hypotheses act as tentative explanations until the research can be carried out to test the hypothesis.

Intelligence Hypothesis (H_1). Nightclubs are an attractive target for takeover by organized crime. A study may have more than one hypothesis, so each hypothesis is labeled H_1, H_2, H_3, and so forth. Then the analyst must construct a null hypothesis (H_0), which is simply the reverse of the research hypothesis.

Intelligence Null Hypothesis (H_0). Nightclubs are not an attractive target for takeover by organized crime. A null hypothesis is established to provide a form of devil's advocate for the analyst. This is because there may be data that falsely support the research hypothesis. So the analyst must aim to reject the null hypothesis. If the null hypothesis cannot be rejected, the analyst can state there is evidence to support the research hypothesis. This is a subtle but important distinction to understand. (See the section entitled "Statistical Significance" in chapter 16 regarding the related issue of type I and type II errors.)

VARIABLES

A variable can take many categorical forms: items, actions, thoughts, mindsets, and moments in time or any number of other category types the analyst is trying to measure. To create a testable hypothesis, the analyst needs an *independent variable* and a *dependent variable*.

The independent variable is independent or not affected by other variables in the study. It acts on the dependent variable, which is the subject of the inquiry. For instance, nightclubs are the independent variable in the example given above about the problem of invasive organized crime.

The dependent variable is, therefore, the factor that depends on other factors—in the current example, organized crime. A simple way of remembering the two variables is to repeat this phrase: the [independent

One technique that is useful for refining a difficult problem is that
of *restating the problem*. Consider the problem of traffic accidents for
the police: how can police officers reduce motor vehicle accidents?
Police may try to do this by increasing the use of radar and mobiliz-
ing high-visibility patrols.

Suppose then that the problem is restated as: how can police
make motor vehicle travel safer? Then solutions that present themselves widen considerably. Additional options could include driver
education programs and working with government departments
that oversee roads and highways to identify dangerous intersec-
tions, turns, or "blind spots." It might also include issuing safety
information to pedestrians, bicyclists, passengers, young drivers,
aged drivers, truck drivers, and motorcyclists.

Restating the problem has the advantage of highlighting different
goals, and these new goals present different solution paths. The
more solution paths there are for a problem, the more likely it is that
a solution can be found to treat it.

variable] causes a change in the [dependent variable]. To illustrate the
point: [nightclubs] cause a change in [organized crime]. Any number of
industries can be substituted for nightclubs to measure if there is any
change in organized crime (i.e., referring back to the research question of
being "an attractive target for takeover by organized crime").

In addition to the distinction between these two variables, it should be
noted that all variables have *attributes* or specific values. The variable gen-
der has two attributes: *male* and *female*.

Moreover, the attributes associated with each variable need to be
exhaustive, so the list should include all possibilities. For instance, if the
variable of religion is used, then limiting the attributes to Catholic, Protes-
tant, Jewish, and Muslim would be a mistake, as there are many more
religions being practiced. One way to address this issue is to expressly
cite the major attributes but then use a category like "other" to account
for the remainder.

Finally, attributes need to be mutually exclusive, so no variable should
have two attributes simultaneously. Even though this may seem self-
evident in practice, achieving this is not so simple. If a crime analyst is
seeking to interview prisoners in jail to gauge their desire to join a "gang,"
then the variable "gang status" with only two attributes, "member" and
"non-member," would be inadequate. How would the analyst record an

answer where a prisoner is a member of a gang but is also seeking membership in a second gang (or wants to quit his present gang and join another)?

OPERATIONALIZING VARIABLES

Operationalizing variables refers to defining the variables in terms so they can be observed and measured. This is necessary because when a phenomenon can be observed, it can be measured, and this measurement can lead to management. But not all variables are easily operationalized. Consider how one would define:

- Organized crime; or
- A nightclub.

Variables that are objective, effort independent (i.e., involuntary), or concrete are easier to observe and, therefore, easier to measure. Those that are subjective, effort dependent, or abstract are much more difficult to operationalize. One way of reducing the confusion (and potential source of error) is to search the literature for other studies that have used the same or similar variables and consider the appropriateness of adopting that definition. Reliability (consistency in measuring) and validity (the degree to which the data collection instrument can measure what is intended) of the operational definition is paramount in the scientific method, as the same results should be obtained if the study is repeated. The process of operationalizing a variable involves three steps:

1. Define the concept to be measured;
2. Assess the best quantitative measures for the concept (as there can be several measures for different concepts); and
3. Consider the most appropriate method for obtaining these data (e.g., covert, unobtrusive, or open source).

MEASURING VARIABLES

There are three methods of measuring variables—directly, indirectly, and through constructs. An example of a directly measurable variable (or an *observable variable*) is the number of people attending a diplomatic function hosted by a target country's government.

An indirectly measurable variable (sometimes termed a hypothetical variable or an indeterminate variable) would be the income of the guests at the diplomatic function. This could be used to assess the position they hold within society.

A variable "observed" through constructs cannot actually be physically observed, as it is an abstract theoretical variable. They are created to represent some phenomenon like the emotional reaction of witnesses to a suicide bomber.

Units of Analysis

An important concept to consider when observing and recording data is the unit of analysis, which is an entity to be analyzed in the course of the intelligence study. A quick way to grasp this is to view an example from the military:

- soldier;
- squad (or section);
- platoon;
- company;
- battalion;
- regiment;
- brigade;
- division;
- corps; and
- army.

It is termed a *unit of analysis* because it is the level at which the study will analyze the data that determines the unit. For instance, if an intelligence study is comparing the type of weapons carried by insurgents who operate from two locations, the unit would be the individual insurgent. However, if the study was comparing how the insurgents engage friendly forces, then the unit of analysis might be at squad or platoon level. Despite this, it is not uncommon for studies to analyze data at several units of analysis.

KEY WORDS AND PHRASES

The key words and phrases associated with this chapter are listed below. Demonstrate your understanding of each by writing a short definition or explanation in one or two sentences.

- Attribute;
- Conclusion;

- Dependent variable;
- Hypothesis;
- Independent variable;
- Null hypothesis;
- Operationalize;
- Premise;
- Probability;
- Proposition;
- Reasoning;
- Scientific methods;
- Significance level;
- Sound argument;
- Source reliability;
- Unit of analysis;
- Valid argument; and
- Variables.

STUDY QUESTIONS

1. Explain why intelligence research employs the scientific method of inquiry.
2. Explain what the two forms of reasoning are and how each could be used in practice.
3. Explain the role probability plays in research.
4. Describe the two integral parts of a properly formed hypothesis and explain why it is constructed in this way.
5. Describe the difference between the independent and dependent variables.

LEARNING ACTIVITY

In intelligence research there is a need to operationalize variables. Consider then, how a crime analyst could define the variable "assaults" so that it can be observed in the community and measured.

NOTES

1. Dr. Edna Reid, Federal Bureau of Investigation, Washington, DC, personal communication, June 7, 2011.
2. Patrick F. Walsh, *Intelligence and Intelligence Analysis* (London: Routledge, 2011).

3. Hank Prunckun, "A Rush to Judgment?: The Origin of the 2001 Australian 'Heroin Drought' and Its Implications for the Future of Drug Law Enforcement," *Global Crime* 7, no. 2 (May 2006): 247–55.

4. Henry Prunckun, *Special Access Required: A Practitioner's Guide to Law Enforcement Intelligence Literature* (Metuchen, NJ: Scarecrow Press, 1990), 3.

5

৯

Intelligence Research Methodologies

This topic provides an overview of the methodologies used in intelligence research by examining:

1. Quantitative research;
2. Qualitative research;
3. Mixed methods research; and
4. Intelligence research designs.

INTRODUCTION

Data are central to intelligence research, whether they are from primary sources, such as field observations, or from secondary sources, like numeric data in its various guises. This chapter examines the three paradigms for collecting, recording, and analyzing data: qualitative research, quantitative research, and mixed methods.

The difference between qualitative and quantitative research is akin to the human body, where quantitative research would be the skeleton, and qualitative research is the flesh. One paradigm complements the other but does not necessarily replace the other. So mixed methods combine both aspects, providing a fuller picture.

Each research paradigm has strengths and limitations. Sometimes the two approaches have been used jointly to *triangulate* research findings (i.e., increase the credibility of the study's results). However, in practice, the use of one over the other often comes down to the researcher's academic background and personal preference. Other times, it is simply a matter of how the research question is framed or what data are available.

QUANTITATIVE RESEARCH

Quantitative research rests with the analyst being able to record observations that can be measured using a measurement *instrument*.[1] These instruments can be self-developed or previously established, as in the case of psychometric tests for vocational aptitude. Some intelligence research projects lend themselves to a quantitative approach where the analyst:

- Is working with large data sets or collecting data from a large number of human subjects (especially if funds and support staff are few or time frames tight);
- Has access to previously developed and tested data collection instruments that are applicable to the project;
- Has a group of decision makers who are more comfortable with numeric data or must demonstrate in measurable terms why a certain recommendation was made; or
- Is intending to estimate or predict some future possible outcome or event based on a sample.

QUALITATIVE RESEARCH

Qualitative intelligence research has the potential to extend understanding by suggesting tentative causal explanations. The analyst approaches the problem under investigation by collecting data in an unstructured way—there are no standardized questionnaires or limiting "boxes to check." Although some qualitative research may use a set of uniform questions to ask each respondent, the answers are not structured in the closed-ended way that is a feature in quantitative research (e.g., characterized by simple "yes" or "no" answers or Likert-scale responses).

Qualitative research can be described as interactive field research or noninteractive documentary research. Data are collected by direct observation in the field or indirectly via diaries, journals, interviews, or focus groups. Secondary sources of data are numerous—documents exist in a myriad of forms. In intelligence research, qualitative data may also form the basis of a pilot study or develop a theory (e.g., grounded theory research) that could later be tested using quantitative data (e.g., using unobtrusive methods).

The keystone to qualitative analysis is *impression*. This is the effect the data have on the researcher who forms a view, image, or opinion about what the data "say." In a practical sense, the analyst examines the data and forms a judgment; then through a process called coding, they describe these impressions as concepts, categories, and properties (*dimensionalizing*). Following on, using one or more analytic techniques, which can include quantitative methods, the analyst then takes these categories and puts them back together in a way that makes connections; it is these connections that produce meaning. Some of the analytics used to make sense of unstructured data are described in detail in chapter 11— "Qualitative Analytics."

An example of using both qualitative and quantitative techniques is in content analysis where the analyst will discuss themes that appear in the text as well as the reading ease and grade level required for adequate understanding. The analyst's impressions of the phenomenon under investigation are usually found in the section of the report that deals with conclusions.

Some intelligence research projects lend themselves to a qualitative approach when the analyst:

- Is interested in the target's behaviors, emotions, or thoughts;
- Has found little in the way of previously published information on the issue, and an overview is needed; or
- Needs an in-depth understanding of the issues (and perhaps associated issues) that quantitative data would not reveal (i.e., there needs to be more "flesh on the bones").

MIXED METHODS RESEARCH

Mixed methods research is a combination of the quantitative and qualitative paradigms. If an analyst is uncomfortable with one or the other, mixed methods is not only a good way of bridging what is sometimes seen as a rift between the two methodological approaches, but also there are indications mixed methods research (sometimes termed *methodological pluralism* or *methodological eclecticism*) frequently results in superior research when compared to mono-method research.[2]

INTELLIGENCE RESEARCH DESIGNS

The most widely used research designs likely to be encountered in intelligence research include the following approaches:

Experimental

Experimental research is used to test for causality where there is the ability to control the variables. In an experiment there may be one or several independent variables that are manipulated to assess their effect on the dependent variable. In this way, the analyst can test whether one or more (or a combination of) independent variables are affecting the dependent variable.

Quasi-experimental

Quasi-experimental research is experimental research designs lacking all the characteristics of a true experiment—notably random selection. Quasi-experimental designs are often in the form of a time-series analysis, which can be either an interrupted time series or a noninterrupted time series (see "Time Series Studies," later in this section). This approach lends itself to analyzing archival and other forms of unobtrusive data.[3]

Case Studies

Case studies are studies of single issues or problems and can be manifested in a person, a group, an incident, or an event. It is a systemic way of examining a problem extending beyond the use of a limited number of variables by providing an in-depth investigation into the target phenomena. Case studies can be single or multiple cases and need not be solely qualitative. Instead, they can use a quantitative paradigm or a mixed approach. This type of research design is well suited to strategic intelligence projects (see, for example, the case study into the 1986 Libyan air raid by the U.S. Air Force in retaliation of terrorist bombings in Europe targeting American interests[4]).

Evaluations

Evaluation research is the systematic assessment of an intelligence operation, a tactical service, or a strategic program. It is usually divided into assessments that are either formative or summative. That is, formative evaluations assess whether the operation, service, or program has been or can be improved through inputs, technology, training, or procedures. Summative evaluations assess whether the desired output or outcome of the operation, service, or program was achieved and if it was not (or partially), suggest improvements it might make to achieve its goals in the future. In strategic intelligence research this is likely to take the form of policy evaluations. So in this sense *outputs* are the units achieved, whereas *outcomes* are the broader police implications. For instance, policy outputs of a new policing strategy might be the number of arrests made by police officers, but the outcome is a lessening of the fear of crime by the community. Evaluations can also study the processes that might be involved in the issue under investigation—known as *process evaluations*. These examine the efficiency and/or effectiveness of the process (note not the effectiveness in terms of outcome).

Focus Groups

Focus groups are used in qualitative research to gather data from a large number of people simultaneously through open-ended questions, although closed questions can be asked to clarify points raised. One strength of this approach is the interactive discussion generated between the participants, creating *information-rich* results. This discussion is usually recorded by the researcher and transcribed later to derive data for the analytic process.

In-Depth Interviews

In-depth interviews are similar to focus groups, although there is only a single or possibly a few respondents. The interview follows an unstructured format; however, the analyst will pose a basic set of open-ended questions to promote discussion and from which the data will be elicited to answer the research question. In-depth interviews are well suited to investigations where personal, sensitive, or confidential (e.g., classified) information is sought, making a group format inappropriate (e.g., counterintelligence investigations).[5]

Ethnographies

Ethnographies seek to answer questions concerning the way people live. Ethnographic research is ideal for examining the relationships between culture and behavior (e.g., insurgencies) and is a good paradigm for gaining insight into an issue that could not be discovered using other research methodologies. While ethnography is a strategic, exploratory method, it can be used in an operational or tactical setting to answer immediately pressing, specific questions.

Grounded Theory

Grounded theory is a research method where theory is developed from the data rather than research conducted to test an established theory. It is a truly inductive approach, taking the analyst from the specific to the general. In grounded theory research, concepts are the important elements of the analysis, as these are used to develop theory (i.e., conceptualization of the data). For instance, a study by Prunckun[6] used this method to develop a theory of counterintelligence. The study used secondary data and extracted key themes relating to counterintelligence practice. Based on these concepts a logical model was constructed that explained the practice of counterintelligence—defensive counterintelligence and offensive counterintelligence. In brief, this theory consists of three supporting axioms and four principles and would have been difficult to achieve without using this approach.

Time Series Studies

Time series studies are sometimes referred to as *repeated measures* studies because there are two or more observations (e.g., measures) taken from the same variable separated by time. Time series analysis aims to understand phenomena represented by a sequence of observations and forecast future values of the variable.

Pre- and Post-Designs

Pre- and post-designs are usually used to measure changes caused by some form of intervention (i.e., a purposefully imposed independent variable on the dependent variable). Measurements are taken before the intervention (baseline data) and then after to assess the causality of the intervention. These types of studies are sometime termed A-B designs, where A represents the baseline phase and B the intervention. Variations to an A-B design are A-B-A, A-B-A-B, and B-A-B. Although it might seem odd to use a B-A-B design, it has its place in cases where the analyst comes across an issue that is in the process of having an intervention applied (B phase), and no baseline data were previously obtained. In such a situation, the analyst can request secession to the intervention in order to measure the post-impact (A) and then reintroduce it (B).

Meta-Analysis

Meta-analysis is a statistical research method that summarizes previously conducted studies. Analysts, rather than survey respondents, survey previously published research reports. It is, therefore, a purely quantitative approach, comparing the statistical results across a number of studies.

KEY WORDS AND PHRASES

The key words and phrases associated with this chapter are listed below. Demonstrate your understanding of each by writing a short definition or explanation in one or two sentences.

- Coding;
- Dimensionalizing;
- Evaluation research;
- Grounded theory research;
- Impressions;
- Information rich;
- Instrument;
- Mixed methods;
- Qualitative research;
- Quantitative research;
- Repeated measures;
- Research designs; and
- Triangulation.

STUDY QUESTIONS

1. Describe the difference between quantitative research and qualitative research.
2. Discuss the strengths and limitations of using quantitative research and qualitative research singularly.
3. Describe how these two approaches can be used to complement each other when used in combination (e.g., mixed methods research).

LEARNING ACTIVITY

Consider the various research designs that can be used in intelligence research. Select one and discuss how this design could be employed to research Country Q, which has recently made it public that it is developing a ballistic missile capability.

NOTES

1. Data collection instruments are sometimes referred to colloquially as *tools*. They are not tools in the true sense, but the term is used as a metaphor to describe

their utility. As useful as this metaphor is, they are better described in a formal piece of writing as *methods*.

2. R. Burke Johnson and Anthony J. Onwuegbuzie, "Mixed Methods Research: A Research Paradigm Whose Time Has Come," *Educational Researcher* 33, no. 7 (2004): 14–26.

3. Richards J. Heuer, ed., *Quantitative Approaches to Political Intelligence: The CIA Experience* (Boulder, CO: Westview Press, 1978).

4. Henry Prunckun and Philip Mohr, "Military Deterrence of International Terrorism: An Evaluation of Operation El Dorado Canyon," *Studies in Conflict and Terrorism* 20, no. 3 (July–September 1997): 267–80.

5. For more information on counterintelligence investigations, see Hank Prunckun, *Counterintelligence Theory and Practice* (Lanham, MD: Rowman & Littlefield, 2012).

6. Hank Prunckun, "A Grounded Theory of Counterintelligence," *American Intelligence Journal* 29, no. 2 (December 2011): 6–15 and Henry Prunckun, "Extending the Theoretical Structure of Intelligence to Counterintelligence," *Salus Journal* (2, no. 2, June 2014: 31–49). See also chapter 3 in Hank Prunckun, *Counterintelligence Theory and Practice*, 2012.

6

🌀

Idea Generation and Conceptualization

This chapter examines some of the most commonly used techniques for generating and conceptualizing ideas including:

1. Brainstorming;
2. Nominal group technique;
3. Mind mapping;
4. Concept mapping;
5. Affinity diagrams;
6. Check sheets;
7. Sorting; and
8. Many-to-many matrix.

BACKGROUND

Filing information consists simply of cataloging data and then storing it for retrieval. Although collation involves these two processes, it also entails some rudimentary forms of analysis. If the analogy of a pole-vaulter is used, the catapulting athlete can be seen as analysis and the run-up to the vault as collation. Analysis cannot be done without collation. And, as with the athlete's run-up to the high bar, how the run-up is executed will see the vault succeed or fail. If collation is done properly, then analysis will be enhanced.

The first phase of the collation process requires the analyst to evaluate the data for inclusion in the study under investigation, or for inclusion in the larger intelligence data holding (e.g., for use in other studies). If the data items are considered not to have value, they should be removed if found to be irrelevant, incorrect, or worthless. Sometimes information that has been collected contains errors or is not relevant. These data should be destroyed after being identified. If they are not, the analyst runs the risk that the information will contribute to congestion in the intelligence database as well as placing the analyst and the agency in legal jeopardy should holding such information be contrary to law.

Data deemed to meet the information collection plan and to be relevant to the project can then be stored for retrieval during the analysis-in-chief stage. Recapping from an earlier chapter, unstructured data is stored electronically by:

- Registering the information;
- Indexing;
- Cross-referencing; and
- Keywording.

This process can aid the analyst in identifying information that has been collected by checking it against the requirements specified in the information collection plan. It also highlights data that were sought but have not been collected. And it can show relationships that hitherto were unknown—especially if the collation technique is visual. As collation techniques are numerous, this chapter will discuss the more widely used techniques used for generating and conceptualizing ideas as they relate to the collation phase of the intelligence cycle.

BRAINSTORMING

Brainstorming has enjoyed widespread use among researchers for decades.[1] Brainstorming is often used in group settings (e.g., say, with all members of the analytic section or a group of subject specialists), and it can be used in a solitary setting by individual analysts.

Although the technique has some limitations, it is arguably one of the most popular methods for collating unstructured data. It can be used as a way to organize data postcollecting, or it can be used in the precollection phase to set up a structure for collecting.

An example of where this technique could be used postcollecting is where field operatives have seized large amounts of data from the target—in the case of a raid on the target's premises—or where an undercover agent hands over large amounts of information surreptitiously obtained from the target. An example of its use in precollection is where an analyst generates a number of ideas for, say, the categories or sources of information that need to be collected in order to answer the research question. This idea generation would be one of the first steps in developing the final information collection plan, which, of course, would include other advice about the collection operation.

Brainstorming is a creative process that results in participants coming up with ideas that are a bit unusual. Nevertheless, it is this feature that allows the generation of numerous ideas in a lateral thinking process. Criticism is to be avoided, and evaluation of the pool of ideas only takes place at the end. Brainstorming sessions need only be short—five to twenty minutes. It is important that the session facilitator does not let criticism creep into the discussion phase and keeps discussion focused on the defined problem (best posted in clear view in front of the group). Encouraging participants to have fun helps generate a wider range of ideas, but balance needs to be struck so that no one vein of thought is perused too far or for too long.

The ideas are recorded so that all participants can see the list, and this helps generate other ideas. If performed as an individual, a mind map can be used instead. If conducted as a group, an affinity diagram can help structure the ideas (see the "Mind Mapping" and "Affinity Diagram" sections, later in this chapter).

But ideas just don't flow from a person's mind like a fountain of creativity. The usual process used to connect the idea at the center of the inquiry with some other idea is through a process of association. For example, while probing possible solutions to why an extremist group might have targeted a particular venue, an analyst might associate it with a hypothesis that she read that addressed a similar issue concerning another extremist group. So in this case association was by "similarity" (i.e., contiguity).

Logic will get you from A to B.
Imagination will take you everywhere.[1]

1. Quote is attributed to the German-born American physicist, the late Dr. Albert Einstein.

Although free association or spontaneous flow is certainly possible, sometimes structured thinking helps. In this regard, some association methods that might be used in a brainstorming session could include thinking about the issue using dissimilarity (contrasts or opposites), historical examples, themes and patterns, random words or phrases, or metaphors.[2]

Some limitations occur when used by groups—participants can become distracted or disengaged mentally because they would rather be doing the work piling up on their own desks back in the office. In addition, some participants may inhibit open and creative discussion if there are personality clashes within the group. Yet still, some may feel self-conscious and may self-censor because they feel they are unable to contribute or that their contribution isn't as worthy as others made by the group. However, the use of brainstorming in groups can have a positive effect that lies outside the production of ideas: that is, it offers a platform for what is termed *buy-in* or *ownership* by those participating.

The 6-3-5 Method

There are several ways to stimulate creative thinking, and the 6-3-5 method is one.[3] Using this method, the analyst convenes a group of six suitable people who are able to consider the brainstorming issue, or several groups of six can be convened simultaneously using this method. Each person is asked to generate three ideas within five minutes, hence the name 6-3-5.

Once this is done, participants pass their sheets of paper to the next person in the group. Each person views the other person's ideas, considers them, and then records three more ideas (within five minutes), thus building on the first three. The process is repeated until each person has had an opportunity to see all sheets of paper.

Variations can be made to this method if time or the number of contributors is constrained, say, a 3-3-2 method, or another variation.

Random Input

Another way to inject a new or different point of view into a problem under study is to use the technique of random input. This can also help regain intellectual momentum when idea generation within the group (or by the individual) has dwindled; or it can be used as a means of obtaining additional ideas by probing additional associations that may not have been considered.[4]

The most straightforward way to do this is to use a dictionary and select a page at random. Then look down the page to the first noun listed. Use this word as a conceptual springboard to make connections between the random word and the issue that is being investigated. This can be repeated several times.

What-If

A brainstorming variation used by fiction writers to generate ideas relating to various elements of story, such as plot and structure, can be adapted in order to collate information relating to intelligence research. This technique is known as the "what-if game."[5] Used in intelligence research, it only requires the analyst to ask "what-if" this-or-that were to occur regarding the issue being considered. Using this method opens up a range of additional possibilities when used as an adjunct to brainstorming.

Devil's Advocate

This can be a useful technique if used correctly during brainstorming. But it can also be a persona that could hold back creativity if used aggressively; its use is a matter of balance and good judgment. Simply, one member of the group acts as the devil's advocate by putting forward opposing ideas, options, or possibilities. It is useful because it steers thinking away from what could be a narrowly focused discussion, thus preventing "groupthink." Groupthink is where members of the group suppress alternative views because they think others may disagree or are embarrassed to voice dissent. Having a devil's advocate in the group can avoid the TINA syndrome (i.e., "there is no alternative") by encouraging members to think wider.

The techniques outlined in the chapter are more than adequate to generate ideas relating to any intelligence research project. Over the years, some creative writers have advocated the use of drugs or alcohol as a means of releasing the mind's creative energies. However, studies have shown that this is not a sensible approach for a variety of physical and mental health reasons. Do not be tempted to resort to drugs or alcohol to "aid" your research. To be the best analyst, you are better off staying healthy in mind, body, and spirit.

NOMINAL GROUP TECHNIQUE

Nominal group techniques can be used as an alternative to brainstorming, especially where the participants may feel some constraint when it comes to articulating their thoughts in front of a group or particular people within the group. The process is ideally suited as a means of circumventing a person(s) who may stifle discussion, because it encourages passive participants.[6] Step by step:

1. Make the team of participants comfortable around a table and supply them with writing materials;
2. Put the question under investigation to the group. It should be
   phrased as an open-ended question in order to encourage thought,
   for example: What are some of the ways Country Q could encourage
   its security forces to do [insert action]?;
3. Request that each participant consider the question for a minute or
   two and then individually brainstorm several possible ideas. The
   brainstorming is to be done silently with the participants' noting
   their ideas on a sheet of paper;
4. Call in the sheets of paper containing the participants' ideas. On a
   flipchart (or via a computer data projector), display the ideas so that
   all can see. During this process, eliminate duplicates by combining
   like with like. As with brainstorming, no criticism is allowed—all
   ideas are to be used. The facilitator can clarify ideas but does so by
   requesting guidance from the group before making any changes;
5. Request that all participants evaluate the ideas by individually (and
   anonymously) voting. This can be done several ways, but two commonly used methods are:

   - Ranking the ideas from most favored at the top to least favored
     at the bottom. Then the rankings are combined and a total cal-
     culated for each idea. The ideas that score the most are the ones
     that deserve attention. Using a scale ranging from 1 to 5 (with
     1 being the lowest and 5 being the highest priorities), table 6.1
     illustrates this. It shows that Issue B would have the highest
     priority (n = 20), with Issue E the lowest (n = 11).
   - The other way is to use a weighting system. For instance, each
     participant can award 100 points across the ideas in any pro-
     portion. This is shown in table 6.2—Issue B has the highest pri-
     ority (n = 165) in this example, with Issue E the lowest (n =
     25).

Table 6.1. Ranking Ideas Method

	Dave	Betty	Kait	Lyndsey	Chris		Total
Issue A	1	5	4	1	2	=	13
Issue B	3	4	5	3	5	=	20
Issue C	3	5	3	4	3	=	18
Issue D	4	2	1	4	1	=	12
Issue E	1	5	2	1	2	=	11

Table 6.2.   Weighting Ideas Method

	Dave	Betty	Kait	Lyndsey	Chris		Total
Issue A		25	40			=	65
Issue B	30	20	50		65	=	165
Issue C	30	30	10	50	35	=	155
Issue D	40			50		=	90
Issue E		25				=	25

MIND MAPPING

A mind map is a diagram representing a set of related ideas. Mind maps can take a variety of forms, but the two most popular are *hierarchical lists* and *spider diagrams*. In practice, it comes down to personal preference of the analyst which to use—some researchers think in terms of lists, while others are visual.

Table 6.3 shows an example of part of a mind map for this book, which was created by the author. It is evident that such a list lends itself ideally to become the book's table of contents (yes, a table of contents is a form of mind map). If an analyst prefers using a spider diagram, it can be converted into a list by grouping (and/or regrouping) the major ideas of the diagram into a hierarchical list. Likewise, a list can be converted into a spider diagram (see figure 6.1).

CONCEPT MAPPING

Concept maps are very similar to mind maps but differ in this regard: where the subthemes are shown in a mind map, concept maps show connections between various idea nodes using descriptive words. In this sense concept maps show the relationship between multiple ideas, and

Table 6.3.   A Hierarchical List of Topics

The Fundamentals of Intelligence	The Intelligence Research Process
Intelligence versus information	Problem formulation
Intelligence defined	Literature review
Intelligence as knowledge	Methodology
Intelligence's consumers	Intelligence collection plan

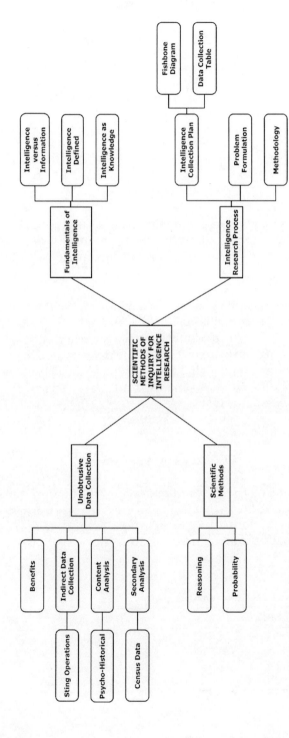

Figure 6.1. Spider diagram of topics.

> "A mind map is the ultimate organizational thinking tool. And it is so simple! . . . Computers can be helpful when you mind map! Although it is still your brain that comes up with all the ideas, the latest software can allow you to draw a mind map on your screen. The advantages of this are obvious. You can save your mind maps in a file and then transmit this information to others. Computer mind maps allow you to store vast amounts of data in mind map form, to cross-reference that data, to shift branches around from one part of the mind map to another, [and] to rearrange entire mind maps in light of new information . . ."[2]
>
> 2. Tony Buzan, *How to Mind Map* (London: Thorsons, 2002), 4, 69.

hence, are useful for depicting concepts diagrammatically. They are particularly useful in investigations where there is some degree of complexity.[7]

As a method of collation, concept maps offer the analyst a way of taking large numbers of related but hitherto unstructured ideas, facts, conditions, attributes, or similar, and collating them so that their relationships become clear. Figure 6.2 shows the beginning of a concept map relating to an intelligence research project on illicit drug importation. The distinguishing descriptive feature—the use of labels for each linking line—helps explain the relationships between the ideas in the nodes. Although only two ideas are represented in this figure, many more can be included, as can be sub-ideas. If a software package is used, it makes linking quicker and it also makes reorganizing or rearranging ideas easier.

AFFINITY DIAGRAMS

Creating an affinity diagram is a method that analysts can use to cluster data items (ideas or concepts) along similar lines—hence, its name *affinity diagram* or *data cluster* (e.g., similar characteristics, qualities, attributes, parts, features, and so on). This clustering should not be confused with classification. In classification, items are assigned according to classes that have been predefined (e.g., in biology or agronomy). In clustering, the objective is to take the data and break it down into smaller clusters, which can be better managed and, therefore, better analyzed.

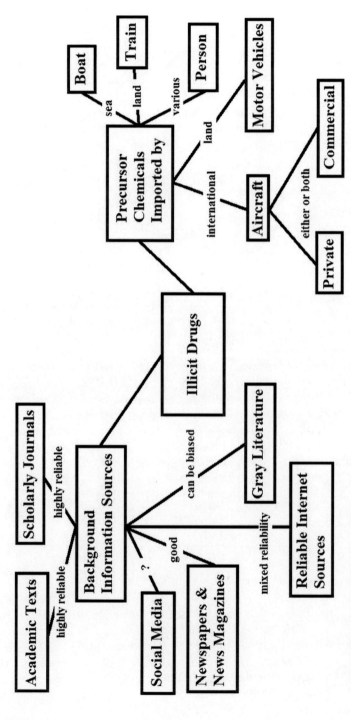

Figure 6.2.   An example of a concept map.

Data clustering can be the follow-on step after a brainstorming session, or it can be used to help sort data that have been acquired by field operatives (e.g., from a search-and-seizure operation or material that has been acquired by an undercover agent or informant). Figure 6.3 shows an example of an affinity diagram of ideas that were generated by an individual who was brainstorming transnational organized crime issues. Step by step:

1. Separate the larger concept/issue/idea/problem into a number of smaller categories;
2. Display these categories via a convenient medium such as a board (e.g., whiteboard, pen board, or marker board), large flipcharts, or computer data projector;
3. Group (and/or regroup) the individual pieces of information (*factors*) into common clusters (as is done in mind mapping); see figure 6.3.
4. Give a heading to each of the clusters (this is done at the end of the process because if assigned initially, the name may change several times as more factors/data items are added, or it may give rise to a newly created cluster when, say, it is discovered that a cluster is too narrow to accommodate all the associated factors).

Organised Crime Groups
- Definitions
- Theories
- History
- Geographic Differences
- Structures
- Functions

Financial Crimes
- Fraud
- Money Laundering
- Counterfeiting
- Securities Swindles
- Stock Manipulation

Smuggling Networks
- Drugs
- Guns
- Illegal Aliens
- Nuclear Material
- Flora and Fauna

Global Impact
- Political
- Social
- Economic
- International Security

Financial Sources
- Trafficking
- Bribery
- Police and Political Corruption
- Motor Vehicle Theft
- Murder
- Extortion
- Prostitution
- Gambling

Control Measures
- Global Cooperation
- Intelligence Sharing
- Cross-jurisdictional Operations
- Extra-legal Operations

Figure 6.3. Affinity diagram of transnational organized crime issues.

CHECK SHEETS

Check sheets aid data collectors and analysts to systematically compile and record data. These data can be either from historical sources or from observations as they occur, and the sheets can be designed and pressed into use very quickly.

The main advantage of using a check sheet system is that as data are entered into the system, a picture is built, making it increasingly clear what the facts are, as opposed to what opinions people are expressing about the problem, issue, or situation.

To create a checklist, the analysts define the events or conditions that need to be observed. They then allocate who will collect the data and specify over what period of time (if it is a phenomenon that is occurring and being observed live) or from what data (e.g., historical records or files). Finally, they then design a checklist sheet (which can be computerized using suitable software) and begin collecting data.

To illustrate this, take the example of a local village that reports seeing an increase in suspicious vehicle traffic passing to its south. It is rumored to be associated with a possible buildup of insurgents in the area. Field operatives are tasked to covertly record this traffic over a twenty-four-hour period using a check sheet. Table 6.4 is an example of such an instrument.

SORTING

Sorting is a simple collation technique, yet this low-level analytical process can yield high-grade results. Let's look at a law enforcement problem as an example.

Assume that a number of conmen have begun selling secondhand vehicles in a major city in your area. These unlicensed dealers pose as private sellers to avoid any liability that would be incurred if they operated under a legitimate dealership. They have rented apartments as their cover for these "private sales" and have placed advertisements in the local newspapers for a number of vehicles.

Now, how can an analyst generate investigative leads for detectives when there are thousands of ads that appear in the classified columns? One way is to sort the raw data. That is, enter the telephone numbers for the various ads into a computerized spreadsheet and then use the "sort" function to either collate the data in ascending or descending order. This will result in a progressive list by number, except where the same telephone number appears more than once. In such cases it would be reasonable to conclude that the person selling these vehicles is an unlicensed

Table 6.4. Example of the Check Sheet Collation Method

Observed	Frequency
Passenger vehicles	~~HHT~~ ~~HHT~~ |||
Light trucks	~~HHT~~ ||
Tractors	~~HHT~~ ~~HHT~~
10-wheeled trucks	||||
18-wheeled trucks	||

dealer. As unlicensed dealers sell multiple vehicles, groupings of phone numbers therefore suggest that these people are worth a visit by a detective.[8]

Although at first glance this may have seemed a difficult investigative task, in the end a simple sort was all that was needed. The effort that was required for the data entry saved far more time in fruitlessly chasing "hunches."

MANY-TO-MANY MATRIX

Analysts can use a matrix diagram to show a *many-to-many relationship* between two data sets. This type of analysis is known as network analysis. In practice, it is a simple but effective technique, as demonstrated in detail in table 11.10 and figure 11.4 of chapter 11 ("Qualitative Analytics"). You will note in these two illustrations that the various relationships are recorded in the matrix, but analysts can convert these data into a network chart (see figure 11.4). Where a relationship exists between the data elements in the matrix, a symbol is inserted. A step-by-step description of how to perform this procedure is provided in chapter 11.

Matrices can also be used to take the collation of information one step further into the realm of analysis. This is done by using a method known as synthesis. For a discussion of this method, see the section entitled "Synthesizing Matrices" in chapter 11.

KEY WORDS AND PHRASES

The key words and phrases associated with this chapter are listed below. Demonstrate your understanding of each by writing a short definition or explanation in one or two sentences.

- 6-3-5 method;
- Affinity diagrams;
- Brainstorming;
- Buy-in;
- Check sheets;
- Concept maps;
- Data clusters;
- Devil's advocate;
- Hierarchical lists;
- Many-to-many relationship;
- Matrix diagrams;
- Mind mapping;
- Nominal group technique; and
- Spider diagrams.

STUDY QUESTIONS

1. Explain the purpose of collation.
2. Describe some of the techniques available to the analyst for collating data.
3. Compare brainstorming with the nominal group technique by giving an example of how an analyst could use each.
4. Could brainstorming and the nominal group technique be substituted for one another? If so, could it be done in all circumstances or only in particular cases? Explain.

LEARNING ACTIVITY

Suppose an analyst is presented with the task of researching retail outlets for counterfeit name-brand goods. Use a mind map to collate information pertaining to the various types of shopping venues that might exist in your jurisdiction and the location of same.

NOTES

1. Alex F. Osborn, *Your Creative Power: How to Use Imagination* (New York: Charles Scribner's Sons, 1948).

2. Alex F. Osborn, *Applied Imagination: Principles and Procedure of Creative Problem-Solving*, third revised edition (New York: Charles Scribner's Sons, 1963).

3. Helmut Schlicksupp, *Kreative Ideenfindung in der Unternehmung: Methoden und Modelle* (Berlin: de Gruyter, 1977).

4. Edward De Bono, *Serious Creativity: Using the Power of Lateral Thinking to Create New Ideas* (Scranton, PA: Harper Business, 1992), 177.

5. James Scott Bell, *Plot and Structure* (Cincinnati, OH: Writer's Digest Books, 2004), 38.

6. A. L. Delbecq and A. H. Van de Ven, "A Group Process Model for Problem Identification and Program Planning," *Journal of Applied Behavioral Science* VII (July–August 1971): 466–91.

7. Joseph D. Novak, *Learning, Creating, and Using Knowledge: Concept Maps as Facilitative Tools in Schools and Corporations*, second edition (New York: Routledge, 2010).

8. Henry Prunckun, "Would You Buy a Used Car from This Man: Utilizing the Intelligence Process in Combating Consumer Fraud in the Second-Hand Motor Vehicle Industry," *Law Enforcement Intelligence Analysis Digest* (Fall 1987: 32–40).

7

⑤

Unobtrusive Data Collection

This topic examines one of the most commonly used techniques for collecting data in intelligence research—unobtrusive methods—by looking at:

1. Benefits of the method; and
2. Indirect data collection techniques.

BENEFITS OF THE METHOD

Unlike surveys, in-depth interviews, and other methods where the researcher has direct communication with the researched, unobtrusive methods attempt to extract data without gaining the target's attention. Unobtrusive methods are conducted without the target knowing about an operation to collect information about him or her.

For intelligence analysts, there are a number of advantages for using unobtrusive methods in their research. Chief among these is what the analyst observes has actually taken place—as opposed to self-reporting by other techniques. These methods are, by nature, intrinsically safe, as they do not place researchers or operations officers (i.e., case officers), agents, or other field operatives (e.g., surveillance teams) in contact with the target or in an environment that could be hostile or dangerous.[1] They are, therefore, discreet and nondisruptive techniques.

Because these methods do not rely on direct contact, they increase the reliability of the study because the research can be replicated, thus allowing checks of the study's reliability and validity to be confirmed. Access does not present a problem because permission is not required, and in the case of a hostile target where permission could not be secured, the analyst does not have to resort to court-approved or executive government–approved covert techniques.

Unobtrusive methods are cost efficient, as undercover operatives do not have to be tasked, nor do surveillance teams or investigators/interrogators have to collect data. Because of its relative economy, unobtrusive methods lend themselves to being used for longitudinal studies where the analyst needs to follow the activities of a target over time.

This, by no means, should underestimate the value of covert and clandestine data collection and data obtained from secret sources, but it highlights the relevance open source information and unobtrusive analytic techniques have in collection plans. In the years since the September 11, 2001 terrorist attacks, the intelligence services have pointed out many times the problems and limitations that the lack of agents on the ground caused for intelligence.[2]

Although unobtrusive methods are discussed here in positive terms (which they are), covert and clandestine sources of information should never be wholly replaced by unobtrusive methods. As discussed in the chapters on quantitative and qualitative approaches to research, one cannot be excluded by the other. In fact, mixed methods research is desirable in many cases, as it increases the reliability and validity of the research findings. The same could be said of mixed data sources—unobtrusive and covert.

INDIRECT DATA COLLECTION TECHNIQUES

Indirect data collection can be used as the analyst's main source of data or used to supplement other forms of data in an attempt to triangulate results. Indirect methods lend themselves to collection by automated and electronic means. The best way to understand indirect methods is by way of example. For instance, if a study wanted to measure the current interest in neo-Nazi issues in a particular region (e.g., if there was some indication of a resurgence in hate crimes), an intelligence operation could set up a website to gather statistical information about those who log on, with the assumption being that those visiting such a website are either ideological supporters of this form of dogma or participants—potential or otherwise—in the ideology's edicts.

Data, such as the country where the Internet visitor originated, could be tracked along with the day, time, and web pages accessed. How long they stay on the website could be measured and whether they were returning viewers or once-only readers. Depending on the construction of the website, many other types of data could be collected, including provision for interactive input by the website visitor. If the analyst in charge of the operation desired, this data collection method could also be used as part of a sting operation where e-mail addresses are obtained via an opt-in facility or a blog.[3]

The study of radio station listening preferences is an example relating to business intelligence that has been cited in the literature from time to time. In this study, the researchers are reported to have conducted an unobtrusive survey of radio stations that were favored by car drivers in a particular geographic area. The researchers attended automotive repair shops, and while customers' cars were being serviced, they noted the current radio station displayed on the radio (whether it was in the AM or FM broadcast band and what particular frequency).

The advantage of this method is that it does not rely on the respondent having to tell the truth. A respondent who admits he or she listens to a radio station that plays music not in popular fashion, or broadcasts certain political messages, or is owned by a particular religious organization, may feel embarrassed about his or her behavior. Prevention of embarrassment may lead respondents to lying, thereby distorting the data. Some of the limitations are that it only captures those respondents who can afford to have their cars serviced at a garage, thus excluding those who service their cars themselves (out of choice or necessity). It also assumes that the current radio station is the one most often or exclusively listened to.

Nevertheless, if a company wanted to market its goods or services to a demographic consisting of Mercedes-Benz owners, then attending garages that repair and service this make of automobile would be a quick and inexpensive way of collecting data about which radio station the business should spend its advertising budget with. It could also be used to verify claims by radio stations about what segment of the market they reach.

This method could also be used to canvass magazine or newspaper readership preferences by scanning the titles of these types of publications that are discarded by residents of a target neighborhood on the day paper products are left at the curbside for recycling. One consideration that needs to be kept in mind when using this technique is that because data are being collected without the target's consent, there may be ethical and legal issues to consider. For instance, with the radio station collection example, one would need to consider whether checking the car's radio without the owner's permission would violate any state or federal statutes regarding privacy.

With the example of inspecting discarded newspapers or magazines, it is unlikely that this data collection would be deemed either a trespass or an invasion of privacy by a court, as these goods, by virtue of their abandonment, are no longer the property of (at that stage) the former owner.[4] However, there may be local bylaws that regulate interfering with material placed on the curbside for rubbish collection or recycling. So before incorporating unobtrusive methods into an information collection plan, seek legal advice.

There are other sources of indirect information, including monitoring radio and television broadcasts for speeches made by political leaders and field operatives attending public speaking forums. In the case of the latter, intelligence analysts need to brief and debrief operatives regarding the specific information required (as per the information collection plan), as well as any additional information gleaned by observation.

More often than not, indirect methods will be appropriate because the data sought are readily available. *Reliability* is an estimate of the *consistency* a collection instrument will yield each time employed or the consistency between two or more sets of data that have been collected.

But just as with other data collection methods, analysts will want to assure themselves of the reliability of the collection method. In this regard, indirect data collection techniques can act as a check for the reliability of another data collection method that may be in use (e.g., a direct method) or for estimating the reliability of itself via a test-retest approach. This is done by collecting data at two different points in time and computing the correlation between the two sets of data. If an analyst considers

Figure 7.1. Paper recycling bins full of potential intelligence data sitting on the curbside outside a business's premises. Note that the bins do not have locks.
Photograph by author

that there is no change in the underlying condition between the two tests, then the data's reliability can be estimated.

KEY WORDS AND PHRASES

The key words and phrases associated with this chapter are listed below. Demonstrate your understanding of each by writing a short definition or explanation in one or two sentences.

- Consistency;
- Indirect data collection;
- Open source information;
- Reliability; and
- Unobtrusive methods.

STUDY QUESTIONS

1. Explain why unobtrusive data collection methods are considered intrinsically safe.
2. Cite three unobtrusive information-gathering techniques.
3. Discuss one way an automated system for collecting indirect data can be set up using the Internet.
4. Outline the main advantages of using unobtrusive methods.

LEARNING ACTIVITY

Suppose that Country Q has had a sudden change of its leadership regime. As such, the agents who were in place previously have lost access to all secret sources of data. Consider how you as an analyst might use unobtrusive data collection methods in the short term to provide the intelligence needed to monitor the situation until operations officers (case officers) reestablish their agent network(s).

NOTES

1. By way of example, see Joby Warrick, *The Triple Agent* (New York: Double-day, 2011).
2. Robert Baer, *See No Evil: The True Story of a Ground Soldier in the CIA's War on Terrorism* (New York: Crown Publishers, 2002); and, Melissa Boyle Mahle,

Denial and Deception: An Insider's View of the CIA from Iran-Contra to 9/11 (New York: Nation Books, 2004).

3. See, for example: Henry Prunckun, "It's Your Money They're After: Sting Operations in Consumer Fraud Investigation," *Police Studies* 11, no. 4 (Winter 1988): 190–94; Henry Prunckun, "Sting Operations in Consumer Fraud Investigation," *Journal of California Law Enforcement* 23, no. 1 (1989): 27–32. See also, Steven K. Frazier, *The Sting Book: A Guide to Setting Up and Running a Clandestine Storefront Sting Operation* (Springfield, IL: Charles C Thomas, 1994).

4. Rick Sarre and Tim Prenzler, *The Law of Private Security in Australia*, second edition (Pyrmont, Australia: Thomson Lawbook, 2009).

8

☙

Open Sources of Information

T his topic briefly examines one of the most commonly used sources for collecting data in intelligence research by looking at:

1. What is open source information;
2. Why open source information;
3. Potential sources;
4. Open versus covert;
5. Benefits of open source information;
6. Libraries;
7. Other sources—where's what;
8. Social media;
9. The Web, and the deep Web;
10. Data mining; and
11. Limitations.

WHAT IS OPEN SOURCE INFORMATION

Open source information is information available to the public.[1] It requires no special authority or needs no special request to be made in order to obtain these data.[2] These sources are in direct contrast to covert and clandestine methods, which are, arguably, the methods most commonly associated with intelligence work. In the parlance of social science research, open source information is categorized as secondary data, and covert and clandestine information are likened to primary data.

Primary and Secondary Data

Primary data is information collected by the intelligence researcher for their specific project. For instance, an analyst may identify in his or her information collection plan photographs of a particular bridge over the Orrenabad River. In this case an operative may be sent to the physical site, and using a cover story, they would obtain the photographs.

In contrast, secondary data is information collected for another purpose (and by others) but usable to answer the intelligence question under investigation. Again, using the Orrenabad River example, secondary data may include commercially available photographs of the bridge from satellites; or photographs of the bridge taken by recent travelers who have posted these on their social media websites; or photographs that related to tourist promotional material (on the Web or in hard copy in libraries); or any number of other publicly available sources.

WHY OPEN SOURCE INFORMATION

Although one might assume that because intelligence research is conducted in secret, the sources of information an analyst uses are also secret. However, as Professor Harry Howe Ransom wrote in his influential work on intelligence: "95 per cent of peacetime intelligence [comes] from open sources."[3] Professor Ransom's own analysis of the United States' national intelligence collection effort stated that in excess of 80 percent came from "overt, above-board methods [that] would normally be available to anyone with a well-organized information gathering system."[4]

The late Richard Helms, former director of central intelligence, stated that between the end of the Second World War when the Office of Strategic Services was deactivated and 1947 when the CIA was created, the agency responsible for secret intelligence was the Strategic Services Unit (SSU). In conducting its intelligence research on the Union of the Soviet Socialist Republics the SSU used the Library of Congress as its main source of data.[5] All the Library of Congress's data in relation to the USSR were publicly available.

With regard to business intelligence, it has been estimated that ". . . 90 percent of all information that you and your business need to make key decisions and to understand your market and competitors is already public or can be systematically developed from public data."[6] As an illustrative point from history, take this Cold War example. Polish intelligence officer Colonel Pawel Monat, who was a military attaché in Washington, DC, saved his Communist government large sums of money, time, and effort by accessing open source information about commercial aviation "secrets." Regarding one experience in particular—that is, with *Aviation Week* magazine—he wrote: "Very little of this information was of really classified nature. We could have dug up most of it ourselves from other sources. But it would have taken us months of work and required us to shell out thousands of dollars to various agents to ferret out the facts, one by one. The magazine handed it all to us on a silver platter [for fifty cents]."[7]

More recently, Tom Clancy was queried about his infallible knowledge of some of the obscure technical and scientific details contained in his espionage novels. He is reported to have denied having access to classified defense information, but instead pointed out what others have discovered—it could all be found in the open source literature.[8]

POTENTIAL SOURCES

Sources of open source information are by no means limited. Open source information can be obtained from newspapers, magazines, academic and professional journals, radio and television broadcasts, and the Internet. Arguably, until the terrorist attacks of September 11, 2001, intelligence analysts used these sources of information simply as a means of supplementing classified information.

> "It is estimated that one weekday edition of today's *New York Times* contains more information than the average person in seventeenth-century England was likely to come across in an entire lifetime."[1]
>
> 1. David Shenk, *Data Smog: Surviving the Information Glut*, revised and updated (New York: HarperCollins, 1997), 25–26.

The importance of systematically collecting and analyzing open source information did not become a priority for the intelligence community until after these attacks. For instance, the effective mining of Internet-based information has enabled the intelligence community to better understand how jihadists use the Internet's Web-television capabilities, chat rooms, and news sites to train their members and raise money.[9]

As a generalization, open source information can be dealt with in an up-front way with the analyst's approaching the source and making a request for the data. This includes approaches such as interviews, surveys, questionnaires, and other forms of direct data collection. But because these latter methods of information collection are adequately covered in the subject literature relating to applied social science research, analysts are referred to this body of information.

OPEN VERSUS COVERT

The prevalence and accessibility of open source information should by no means underestimate the value of covert and clandestine data collection and data obtained from confidential sources. It simply highlights the relevance open source information and unobtrusive analytic techniques have in collection plans.

Open sources, used astutely, can be a positive boon to information collected by covert and clandestine means. But as pointed out with regard to unobtrusive methods in the previous chapter, one should not replace the other.[10]

BENEFITS OF OPEN SOURCE INFORMATION

One of the central concerns for covert data collection is that of safety to the officer or agent tasked with collecting information. The risks are real and the consequences of failure are grave—both to the operative and the employing agency.

However, open source data collection does not pose the same risks, and the depth and breadth of the information is potentially vast. Having said that, the latter point can also be a drawback—oftentimes if a specific piece of information is needed, it may be classified, and hence an agent working undercover may be the only way to gather it. In this regard volumes of information cannot be substituted for specificity.

Nevertheless, open source information can facilitate a quick means of supplying data to answer the research question and therefore assess a developing situation. It can also be used to determine the missing pieces of information (i.e., the *specificity* just discussed) by highlighting what is required in an information collection plan. Such a plan may at this stage point to the need to involve covert means.

LIBRARIES

A sometimes overlooked source of open source information is the public and university library. Analysts should never overlook the time-honored library. This is a rich source of data. Location-dependent, the quality and quantity of a library's holdings will vary, but generally, libraries have extensive holdings of nonfiction reference works as well as access to the interlibrary loan program. The latter provides access to nationwide holdings through member libraries.

In addition to the wealth of information contained in the library's stacks, libraries have reference books, maps, newspapers, journals, periodicals, registers, and catalogs. There are also special collections within libraries and privately maintained computer databases that contain topics of particular interest. An example of the former is the Russell J. Bowen Collection of Works on Intelligence, Security, and Covert Activities, which is housed in the Lauinger Library, Georgetown University, Washington, DC (see figure 8.1).

Prior to digital collections, searching and retrieval of information from a library was a time-consuming task. It sometimes required the analyst to search hard-copy index cards, cross check the desired entry with another list of, say, journals held by the library, and then locate and copy the article. If the journal was not held by the library, then the process of acquiring it via internal loan or physically visiting another library would be the only option.

However, digital collections have eliminated these inefficiencies. Now an analyst can do a search of a library that may not even be in his or her state or country from his or her computer workstation. From that search, the analyst can locate a journal article and then download it in a readable format that looks exactly like the printed form in the hard-copy journal. Many books can also be downloaded in electronic form. As an aside, a growing number of analysts have degrees in library science, as the skill sets for both professions share commonalities.[11]

Private Professional Collections

Analysts can put together their own personal special collection of intelligence reference books too. This can be done relatively inexpensively by visiting secondhand bookstores in their area. There are often titles relating to many areas of interest to scholar-spies—research methodologies, analysis, statistics, as well as a range of topics within intelligence and counterintelligence (e.g., cryptanalysis, espionage tradecraft, history of intelligence, spy memoirs, investigative exposés, etc.). Analysts can also buy books on subject areas that they might be focusing on in their investigations—arms control, weapons proliferation, transnational crime, terrorism, arms and drug trafficking, organized crime, corruption, cyber-crime, war crimes—and the list could go on and on.

If analysts are located near a college or university, they might be able to find secondhand bookstores that have taken in last semester's student texts. These will have a potentially richer source of book titles. Although buying books this way is largely a hit-or-miss affair, it is an inexpensive method of acquiring a professional library at a fraction of the cost of buying new books. And it can be fun hunting through the shelves each month

Figure 8.1. Some of the texts on secret intelligence held by the Russell J. Bowen Collection of Works on Intelligence, Security, and Covert Activities at Georgetown University.

Photograph by author

to see what new "treasures" might be found. There is nothing like having a room with floor-to-ceiling and wall-to-wall bookshelves full of books.

If analysts do purchase books to create themselves a private professional collection, it is recommended that, where possible, they buy hardbound editions. These will last longer as they are more durable, especially as the years pass. Some private professional libraries have hardbound intelligence-related books dating back to the First World War that were bought for as little as two dollars in secondhand shops. These texts have not been seen in any other special collections. The point is that books do not have to be the latest release to be of value—historical information is continuously of value in intelligence research (recall our discussion of "basic intelligence" in chapter 2).

"There are worse crimes than burning books. One of them is not reading them."[2]

2. Attributed to the Russian-born American poet and essayist Iosif Aleksandrovich "Joseph" Brodsky (1940–1996). Brodsky was the winner of the 1987 Nobel Prize in Literature.

OTHER SOURCES—WHERE'S WHAT

In the mid-1970s the Central Intelligence Agency's Office of Security prepared a book outlining sources of information for federal investigators. The guide was entitled *Where's What*[12] and documented thousands of sources of information. Many of these are publicly available, but others can only be accessed by government employees with an appropriate authority. Even though the reference is somewhat dated, the sources remain relevant now as they did then (though, perhaps, the sources would now appear under a different departmental name due to restructuring and corporate stylizing).

There are numerous sources of information in the form of professional and academic documents. For example, analysts can access conferences and symposia papers, information published by professional associations, as well as college and university professors and subject matter experts in industry.[13]

The United States has its government printer, the Library of Congress, *Congressional Record*, National Archives, and information available through the Freedom of Information Act. In most liberal democracies, there exist equivalents to the freedom of information legislation as well as equivalents to the holdings of the U.S. archives cited.

Finally, there are public radio and television broadcasts, which have the potential to provide an extremely wide range of information, as do photographic and motion picture archives. Many of these sources are now available in electronic form and can be downloaded in formats that can be played or viewed on the analyst's computer workstation.

Some examples of less-thought-about sources of open source information can also include telephone directories (e.g., backdated); city directories; vehicle license plates; drivers' licenses; birth, death, and marriage records; civil and criminal court records; property titles, mortgage documents, liens, and caveats; school records; voter registration lists; credit reporting agencies; utility companies; credit card companies; insurance companies; stockbrokers; moving companies; chambers of commerce; racing or gaming commissions; banks and finance companies; the postal authority; numerous government departments, agencies, and statutory bodies (local, state, and national); and employment agencies.

Like library materials, some of these sources are now available online, but for historical research, a trip to the library to view, say, an old telephone directory may be essential. Although one might question this, consider a research project where an analyst is probing the *bona fides* of, perhaps, a walk-in defector to see if he is genuine, or whether his past was the result of fabrication through the creation of a cover, or *legend*.

Gray Literature

Generally speaking, much of the material discussed so far finds its way to the analyst's desk via one or more of the conventional avenues for publishing. These publishing methods are usually associated with a commercial publishing house of some kind—usually a well-recognized publisher.

There is another avenue where information gets presented to the public domain, and this is termed in the library sciences as *gray literature*. It comprises information that is published in an informal manner that can vary in format—from printed to electronic. The gray literature is usually associated with technical reports, scientific papers, working papers and notes, committee reports, white papers, and academic preprints.

These "publications" originate from a diverse range of sources—for example, university and college research centers, think tanks, consulting firms, private researchers, government bodies, nongovernment organizations, and commercial businesses. There has been an exponential increase in the number of these gray publications as a result of the ubiquity of digital technology.

Master's theses and doctoral dissertations are not considered part of the gray literature because, although they are prepared for a form of publication, they are prepared for examiners, who perform a form of peer review of the research. Although the gray literature presents a rich source of information for analysts, it should be noted that these publications are distributed outside of the peer-review process and hence should be viewed with caution until each is assessed for validity and reliability.

As a guide, there are three areas where analysts should direct their focus when triaging gray publications: (1) the nature of the document; (2) how the document is distributed; and (3) the source of the document. With regard to the latter, the source of funding for the research presented in the publication may play an influencing factor as to the objectivity of the material. Questions the analyst can ask of the material include: Where does the sponsor's funding come from? Has this been made transparent in the document? And has the funding been provided to benefit some organization in particular?

Users of gray literature are very complimentary of this sometimes overlooked and underutilized source, but equally users point out the limitations and stress the need to evaluate every document before placing any weight on its contents. It has been reported that some gray literature is "counterfeit"—that is, there are documents that are self-serving and masquerade as gray literature. These documents promote a particular point of view under the guise of being an authentic contribution to knowledge.

Some sources of such bogus documents come from industry groups, commercial firms, nongovernment agencies, and think tanks, as well as others. This is not to say all of these types of groups are involved in such questionable publication practices. Far from it, but nonetheless, users of the gray literature have reported that there are publications that present the pretense of academic rigor, but their publication is merely part of a media campaign to promote a "political" message or to increase a company's profit or organization's standing.

As a final thought about gray literature—it is a useful supplement to the peer-reviewed subject literature but should never be a substitute for it.

SOCIAL MEDIA

Paper Trails

Because we live in an age where information is central to every aspect of life, an astonishing variety of obligations have been placed on individuals to record information about their affairs. The same applies to corporate bodies and governments. It is because of these record systems that society generates what has become known as a *paper trail*.

A paper trail can be described as all records and documents created by an individual or entity in the course of commercial and social interaction with other individuals, organizations, government departments, and businesses (both public and private). These records and documents have the effect of leaving a trail detailing where the individual (or entity) has been, with whom they have had dealings, what goods and services they have purchased, what they own, what their likes and dislikes are, and more importantly, what their intentions may be.

To the analyst, this trail forms a composite picture for any target that comes under surveillance. Uncovering one part of the paper trail can lead the analyst to other sources of information. These sources are not only limited to those on paper but also can be extended to interviews with friends, neighbors, and colleagues of the target, perhaps using a pretext as well as physical, optical, or electronic surveillance.

To the analyst, the paper trail is a valuable set of leads. The value of this information in a collated and analyzed form can be seen in the arrests reported in the media from time to time of influential organized crime members that were, until that point, untouchable by law enforcement agencies.

But a paper trail does not need to be on paper. With social media, what used to be recorded on paper now appears in electronic form. Any form of electronic media where people record social interaction and are accessible to others satisfies this definition. Information about a person's ideology; interests; movements; past, present, and future activities; education; employment; and so on, form a long list of the types of data these media record. But unlike the paper trails of the past, these electronic paper trails also hold audio files, video files, and photographs.

Application for Intelligence

The use of social media for intelligence has great application. As people's lives, and the lives of many organizations, are captured every day and in many ways using social media, these data can be incorporated in information collection plans. These data can be collected, collated, stored, analyzed, and shared between intelligence analysts and agencies. Data harvested via online search engines can be done without the knowledge of the person to whom the data relate.

Social networking websites can be used to conduct background investigations on people. Take, for instance, the situation where an operations officer plans to pitch to a potential agent at a public event (e.g., the opening of an artist's recent paintings). An analyst could therefore compile a background file on the potential agent using social media, including social networking, in order to provide the operations officer with information so they can strike up a conversation with the potential agent.[14]

It would be fair to say that there are few people who do not resent the thought of being surveilled by their government. The thought of a Big Brother society where the state conducts surveillance on everyone conjures up notions of shadowy, sinister, secretive, and repressive regimes. But people around the world now collude with this type of surveillance by willingly posting personal information to social media outlets.

In some regards today's world may have astonished and disturbed George Orwell,[15] but social surveillance is accepted without concern by the majority of the population, and as such can be and is exploited by intelligence practitioners. Take as an example the notorious Jordanian terrorist Humam Khalil Abu Mulal al-Balawi, who began his activities in chat rooms on social media websites. It wasn't long before the world's intelligence agencies began monitoring him through these media and compiling a dossier on him and his activities.[16]

Research suggests that there is a type of person known as a "street stroller." This type of person is said to enjoy being noticed in public, so they openly display themselves—like actors on a stage. If this concept is translated to the world of social media, it could be said that these street strollers have become "cyber strollers." And as such, it would pay dividends for analysts to include these sources of information as part of any information collection plan.

Limitations and Unintended Consequences

As rich as this source is for information, social media has limitations. One example that demonstrates this is the use of *sock puppets*. A sock puppet is a false identity that is created as a deception so that the creator can manipulate the ideas posted on social media websites.[17] This is different from the use of a pseudonym because a sock puppet purports to be another person—a real person, but no such person exists. The person is computer generated with algorithms crafted by the third party. These algorithms are usually based on narrative and statistical probability, as well as other methodologies. Software exists that allows the simultaneous control of numerous false identities.

If used by an intelligence agency, "online operations officers" (like case officers in field operations) can control these "fake agents" for the purposes of, say, penetrating extremist social media forums and sowing information that would lead to an intelligence project's objectives being realized. Such objectives could be wide-ranging—from simple disruptive activities aimed at a targeted online community to influencing future actions of these people in the same vein as would an agent provocateur.

This is done by what has been described as personality-based social-Web robots (or *bots*). These robots use keywords that define the persona of the fake agent to automate the posting of computer-generated blogs on all of the various types of social media that exist.

The implication for analysts who are collecting information from social media is to be mindful that colleagues in their agencies, or allied organizations, could be running sock puppet agents online and they should not mistake this information for genuine information posted by the target group. Likewise, analysts need to be conscious that opposition agents may be doing the same and that such websites may contain misinformation and/or disinformation (see chapter 3 about the issues involved in data evaluation). Further, social activists may be using these techniques, as well as those developed by marketers, to exploit the fundamental flaws in online journalism to manipulate public opinion.[18]

As social media websites are accessible not only by individuals and groups of the target population who may reside outside of the geographic bounds of the analyst's country, citizens of his or her country may also be accessing/participating in these online discussions. As such, another unintended consequence is that the manipulation of this online information may affect public opinion. This issue has consequences in liberal democracies, and it should not be overlooked or discounted. History shows us the ramifications of government officials who have not heeded this lesson. . . . There may also be legal implications in terms of the perennial issue raised by defense counsel—in jurisprudence, it is known as *entrapment*.

A variation of automated creation of fake social media content is the employment of what has come to be called *crowdsourcing*. This is where very large numbers of people, usually in developing economies, are employed to create fake accounts on social media websites and post biased information that supports the "employer's" aims. This could be to add praise to the goals associated with the employer or degrade the opposition, or both. This technique can be used by commercial firms selling goods or services, or security agencies of various nations. In any case, this can be done under the auspices of a "false flag" cover, or by way of a company or organization that is set up in order to isolate the employer, thereby affording deniability. For example, along with personality-based social-Web robots, this technique could be very effective in influencing political change in countries that are the target of intelligence operations.

Crowdsourced Investigations

The April 15, 2013, Boston terrorist bombings led to a flood of photographs submitted to law enforcement agencies by civic-minded people.

Photographs were forwarded to the FBI by people who attended the Boston Marathon and answered the FBI's call for assistance.[19]

Photos taken before, during, and after the attack ultimately helped identify two suspects. However providing these photographs gave rise to people using these images to conduct unprofessional investigations online. Although well intentioned, these substandard investigations resulted in the identification of people who had nothing to do with the bombing—innocent people. Counterterrorism expert Associate Professor Nicholas O'Brien of Charles Sturt University stated: ". . . the quality of the analysis was decidedly hit-and-miss—a heady mix of 'trolling,' genuine analysis and well-meaning, but ultimately unhelpful amateurism."[20] Reading some of the posts on these websites showed these unprofessional sleuths engaged in speculation that could only be described as precarious when looked at in the context of intelligence research which is based on the scientific method of inquiry.

No doubt the response from the public for images of such events should they happen again in the future will be a feature. As such, intelligence analysts need to consider systems to store, collate, retrieve, and analyze the large volume of photographic data that they will receive. But also, they need to consider another aspect—willful sabotage by people who may or may not be associated with the perpetrators. This might include disgruntled people submitting altered images to throw the investigation off track or waste analysts' and investigators' time. Or these images may be spitefully embedded with viruses designed to infect the agency's data processing systems. But they also need to realize that they will be competing with untrained and unskilled online "detectives" who, despite their good intentions, will be confounding their inquiries and perhaps fueling misinformation.[21]

THE WEB, AND THE DEEP WEB

The World Wide Web, or Web for short, has been in existence since 1991. Its purpose is to provide access to information via data stored on servers around the world. Estimates vary because information on the Web is being added, removed, and changed daily, but as an indicative number, in 2010 it was estimated that the *surface Web* contained about six hundred billion pages of information (i.e., 600,000,000,000).[22] All indications suggest that although material is removed from the Web, exponentially more is being added. To access this information, people use browser software and conduct their search via a standard search engine, which has indexed these data.

However, not all information is indexed by these largely commercial search engines. That is because the criteria and indexing algorithms do not capture all the information that exists on the Web. This untapped part of the Web is termed the *invisible Web*, *hidden Web*, or the *dark Web*. Regardless of the term used, they refer to this enormous uncaptured storehouse of online information.

These data are buried in databases and other research holdings on servers around the world. Although they are technically searchable and hence accessible online, they are usually omitted by standard search engines. Searching the deep Web is done through a tailored search interface.

Like the surface Web, estimates of what is contained in the deep Web vary and are based on extrapolations. Some have claimed that it is 550 times larger than the surface Web[23] and contains tens of thousands of terabytes of data. To the intelligence analyst, this massive amount of information represents a potent source of easily accessible information.

Limitations and Unintended Consequences

The thought of searching for intelligence material on the Web or the deep Web might seem, at first glance, a relatively harmless task with little risk, but there are two levels of concern that analysts should bear in mind.

The first is that a visit to a website relating to any issue of concern for a law enforcement or intelligence agency is likely to be monitored by an opposition agency. In the physical world, think of a "trip wire" that triggers an alert once someone steps past a certain point. In cyberspace security and intelligence agencies are able to install similar trip wires that log the IP address[24] of people visiting certain sites, or searching for or downloading information. (Fake websites may also be set up by intelligence agencies as decoys to track people interested in issues of concern to them.)

The trip wire algorithm used to alert analysts will vary from agency to agency and from issue to issue, but suffice it to say that searching for material on jihadism or downloading material issued by jihadist groups are likely to trip an alert somewhere with a law enforcement–type agency. There could be other keywords or phrases—for instance, bombs and explosives, weapons, nuclear material, drugs, money transfers, or certain location names, or certain people's names, and so on. If the research project is classified, then searching for or downloading material is likely to leave a paper trail that leads some interested opposition agency to the analyst's desk. If that agency is not an ally, this is a problem. Even if the agency is a friend, it may not have a need-to-know or there may be no reason to share this information with them; likewise, this is a problem.

The second level of concern is related to trip wires placed by the targets of intelligence investigations. Like a security and intelligence agency that logs searches and downloads, the targets themselves can log IP addresses and other access information about those who visit their sites. Therefore, an analyst may inadvertently tip off their target if some counterintelligence strategy is not employed.[25]

DATA MINING

The term *data mining* conjures up the notion of an analyst "digging" into a number of data sets to extract "nuggets" of informational gold. In a way, this is what the process and practice of data mining is about. At its core, an analyst uses software applications to interrogate a number of rational databases in order to discover correlations or patterns in the data that will eventually lead to insightful conclusions being drawn.

Data mining is not new—it has been around for decades. However, it was expensive to perform and the data sets used were, generally, smaller than those available at the time this edition went to print. The computing hardware was also expensive and less powerful, as were the software applications used to interrogate the databases.

Hardware and software developments, along with data storage devices, have increased processing speed (e.g., multiple processors), sophistication, and capacity, respectively. Moreover, the cost of purchasing and running data mining systems has dropped considerably from its early days to the point where private researchers and contract analysts can afford these systems. Databases comprising many terabytes of information (known as *data warehouses*) can be accessed for a fee, so analysts do not have to be involved in the collection aspects of the process (which can be considerable).[26] The data sets themselves can also consist of structured and unstructured data, which was not always the case.

"Simply stated, data mining refers to *extracting or 'mining' knowledge from large amounts of data*. The term is actually a misnomer. Remember that the mining of gold from rocks or sand is referred to as *gold* mining rather than rock or sand mining. Thus, data mining should have been more appropriately named 'knowledge mining . . .'."[3]

3. Jiawei Han and Micheline Kamber, *Data Mining: Concepts and Techniques*, second edition (Burlington, MA: Morgan Kaufmann, 2006), 5.

It would be fair to say that originally data mining was used by the business sector to assist its market research. However, once intelligence agencies comprehended the value of this type of analysis, they embraced it, and several large-scale data mining projects were operating post-9/11. Nevertheless, concerns about citizens' privacy and civil liberties soon became a feature of debate because along with data about potential targets were data about law-abiding people who live respectable lives. There were also security concerns about inadvertent access or deliberate "hacking" of the data mining systems by third parties as well as the potential penetration by opposition intelligence services.

The value of data mining is to take disparate data sets and with a few well-designed software applications, allow analysts to develop complex queries so they can interrogate these databases to see if certain relationships emerge. These patterns are based on association rules which are grounded in logic and/or mathematical concepts. In a sense, these queries can be likened to hypotheses that are tested against the data to see if they hold true.

Once relationships start to develop, other software applications can be employed to graphically display the results and allow secondary queries to be run. Like finding nuggets of pure gold in a ton of rocks, data mining allows terabytes of information to be processed to find relationships that would be impossible if performed manually (if they could be performed manually, which is not likely).

What does data mining hold for intelligence analysis? Arguably two key strategies are used by intelligence analysts, and these are pattern mining and subject-based mining. With the former, analysts query relationships between variables in the data sets to discover patterns. Because the databases are relational, that means many more variables can be created by defining them in terms of the existing data items (known as *derived data*). For instance, a new variable may be created by defining it as a data item where attributes A and D were present, but only if these attributes occurred in time before attribute M. By defining new variables this way, data items from many different relational databases can be queried to produce these new variables. These new variables can then be queried in regard to other existing variables or other new derived data items.

With regard to subject-based mining, the analyst starts with subject-specific information items and then mines the data warehouse at his or her disposal to create a profile or dossier of related information to the initiating data. *Profile* and *dossier* are used here to denote a collection of information that may be either narrow or wide ranging, relating to many aspects of the target's life.

Being unchained from having to conduct individual queries on stand-alone small data sets that are characterized by predefined search options is a very powerful method for analysts. When this methodology is coupled with multiple relational databases containing terabytes of information that can be interrogated by sophisticated software using computers with several multicore processors (and distributed across several such servers), data mining becomes the material that Hollywood movies are made of.

By way of example, if a government analyst was looking for a terrorist cell, he or she might, in theory, query a number of government databases by using existing data items (variables) along with derived data items. These might also be combined with business-related data that is accessible for a fee from the commercial sector. For instance, when considering just these common databases,[27] one can appreciate the power this methodology wields when trying to track down a target, and one can appreciate why civil libertarians recoil with fear about the method's potential misuse:

- address records;
- aircraft registrations;
- airline manifests;
- birth records;
- boat registrations;
- civil court records;
- company and business registrations;
- credit card purchase data;
- credit histories;
- divorce records;
- driver's license information;
- freight records;
- genealogical/family history databases;
- geospatial data;
- hotel/motel reservations data;
- land ownership records;
- marriage records;
- money movement/bank transfer records;
- motor vehicle registrations;
- newspaper archives;
- occupational licensing records;
- police criminal records;
- shipping records;
- Social Security records; and
- social Web, including blogs.

Analytic Software

If an information collection plan includes large data sets that are characteristic of data mining operations, then manual procedures become unrealistic. Analytic software must be used to be able to filter, sort, and collate the raw data, and then to present the findings of various queries. However, it is important that analysts have an understanding of research methodologies as well as the nature of the data they are using, along with its limitations. Without this understanding the analyst is not an analyst, but merely a trained collator. There is nothing wrong with having staff trained in the operation of such software programs, but without an understanding of the underlying theory and practice of intelligence research, the risk of errors increases. For instance, incorrect or overextended conclusions could be drawn from the results of queries; "outliers" mistaken for part of a pattern of results; queries could be run against incompatible data sets; or type I or type II errors could be overlooked or not recognized. In short, a wide range of incorrect results could be produced based on a poor or partial understanding of the scientific method of inquiry as it applies to intelligence research.

There are many analytic software packages on the market, and most are very sophisticated, delivering wonderful visual products for briefings and reports. Hollywood movies and television shows have seized on the sophistication of these products, and made-for-cinema variants show up in crime and espionage thrillers regularly.

The developers of these software programs license individual workstations as well as server-based applications that run over a variety of networks. Training in how to use these applications is usually provided by the software supplier, though there are many textbooks written by third-party authors about how to use these programs.

Although these products are usually "easy" to use, there are many functions and features, so to become proficient in all aspects of the software a short training course is required. Some operators of these software programs therefore confuse this software training with intelligence training—it is not the same. Without demeaning the importance of the role these *computer operators* perform in intelligence agencies or their skills, this training could be said to be analogous to the training undergone by a butcher when compared to the education undertaken by a surgeon. Given the weight resting on the judgments formed by analysts, it is a compelling argument that an analyst requires at least a bachelor's-level degree, but preferably a master's or doctoral-level qualification. On-the-job training is important to operate these software applications efficiently, but this training is not sufficient to qualify one to become an analyst. *Training* should never be confused with *education*.

LIMITATIONS

If open source information is information available to the public and requires no special operational agents or teams of operatives to collect it;

if its collection is not dependent upon a court-ordered warrant or other form of legal authority; if it does not require large operational budgets to support the logistical infrastructure that covert methods require; then one could conclude it is approaching a flawless source of information.

But this is not the case. Open source data collection has limitations just as covert and clandestine methods do. Although a category of every possible limitation is not possible here, suffice it to say that intelligence analysts who use open source information need to be mindful of the same issues that are present in obtaining information from other sources. In particular, it is worth noting that because of the nature of open source data, analysts should be particularly concerned with deception and bias. This is because unlike methods that collect information through some form of secret observation, secondary sources of data are susceptible to intentional as well as unintentional changes. "It is very important to know the background of open sources and the purpose of the public information in order to distinguish objective, factual information from information that lacks merit, contains bias, or is an effort to deceive the reader."[28]

KEY WORDS AND PHRASES

The key words and phrases associated with this chapter are listed below. Demonstrate your understanding of each by writing a short definition or explanation in one or two sentences.

- Data mining;
- Data warehouse;
- Deep Web;
- Derived data;
- Legend;
- Open source information;
- Paper trail;
- Sock puppets; and
- Surface Web.

STUDY QUESTIONS

1. Explain why intelligence analysts should consider the use of open source information in their collection plans.
2. Discuss how the use of open source information might reduce the risk of collecting information using an undercover agent.

3. List at least twelve open sources of information that might be of value to an analyst that are contained in a library.
4. Describe what a *paper trail* is and how this concept is projected into the digital world of the Internet.
5. Discuss why it is difficult to estimate the amount of information contained in the Web—regardless of whether it is the surface Web or the deep Web.

LEARNING ACTIVITY

Research the different types of tailored search interfaces that are available free on the Internet to query the deep Web. Using one of these facilities, search the Web for information about one of these issues: (1) arms trafficking; (2) Asian organized crime; (3) jihadist ideology; (4) radicalization; or (5) cyber weapons. Then conduct this same search using the surface Web through one of the standard search engines. Compare the findings of the two searches. What differences were noted? From the point of view of an analyst conducting a research project on the issue you selected, argue whether using one or the other, or a combination of both types of searches, would yield the most benefit.

NOTES

1. Sometimes referred to as open source intelligence and abbreviated OSINT.
2. Although for some agencies, in particular the military and national security agencies, there may be restriction imposed by regulations or directives that prohibit the collection, retention, or dissemination of information regarding U.S. citizens. See, for instance, Army Regulation 381-10, *U.S. Army Intelligence Activities*, and Executive Order 12333, *U.S. Intelligence Activities*.
3. Harry Howe Ransom, *The Intelligence Establishment* (Cambridge, MA: Harvard University Press, 1971), 19. Professor Ransom was quoting Ellis M. Zacharias, a World War II deputy director of the Office of Naval Intelligence. According to Zacharias, only 4 percent of intelligence came from semi-open sources, and a mere 1 percent from secret agents.
4. Ransom, *The Intelligence Establishment*, 20.
5. Richard Helms with William Hood, *A Look Over My Shoulder: A Life in the Central Intelligence Agency* (New York: Random House, 2003), 73.
6. John J. McGonagle Jr. and Carolyn M. Vella, *Outsmarting the Competition: Practical Approaches to Finding and Using Competitive Information* (Naperville, IL: Sourcebooks, 1990), 4.
7. Pawel Monat with John Dille, *Spy in the U.S.* (London: Frederick Muller Limited, 1962), 120.

8. Frederick P. Hitz, *The Great Game: The Myth and Reality of Espionage* (New York: Alfred A. Knopf, 2004), 86.

9. Richard Best and Alfred Cummings, *Open Source Intelligence Issues for Congress* (Washington, DC: Congressional Research Service, 2007).

10. Robert Baer, *See No Evil: The True Story of a Ground Soldier in the CIA's War on Terrorism* (New York: Crown Publishers, 2002); and Melissa Boyle Mahle, *Denial and Deception: An Insider's View of the CIA from Iran-Contra to 9/11* (New York: Nation Books, 2004).

11. Dr. Edna Reid, Federal Bureau of Investigation, Washington, DC, personal communication, June 7, 2011.

12. Harry J. Murphy, Office of Security, Central Intelligence Agency, *Where's What: Sources of Information for Federal Investigators* (New York: Quadrangle/The New York Times Book Co., 1975).

13. Mark M. Lowenthal, *Intelligence: From Secrets to Policy*, second edition (Washington, DC: CQ Press, 2003), 79.

14. In the counterintelligence context, access to these types of data via social media is a reason why intelligence officers should avoid or restrict the personal information they post to the Web. However, there is a twist to this if the officer is an undercover operative, as having a social media presence may be needed to establish a cover or legend.

15. George Orwell was the pseudonym for Eric Arthur Blair, who wrote the fictional novel *Nineteen Eighty-Four*. In this book Orwell portrays a fictional totalitarian society where government surveillance is omnipresent. Although he describes the system of surveillance that consisted of agents, informants, and two-way "telescreens," the computer monitor and social media could, by analogy, be considered as a manifestation of the latter. George Orwell, *Nineteen Eighty-Four, A Novel* (New York: Harcourt, Brace, 1949).

16. Joby Warrick, *The Triple Agent: The al-Qaeda Mole Who Infiltrated the CIA* (New York: Doubleday, 2011).

17. See also accounts about how marketers have exploited the online journalism to successfully manipulate public opinion. Ryan Holiday, *Trust Me I'm Lying: Confessions of a Media Manipulator* (New York: Portfolio/Penguin, 2012).

18. Holiday, *Trust Me I'm Lying: Confessions of a Media Manipulator*.

19. For example, via the FBI's website that was established for this purpose: https://bostonmarathontips.fbi.gov/ (accessed April 23, 2013).

20. Nicholas O'Brien, "Did citizen sleuths give the FBI a run for its money in Boston? No," *The Conversation*, https://theconversation.com/did-citizen-sleuths -give-the-fbi-a-run-for-its-money-in-boston-no-13663 (accessed April 23, 2013).

21. See chapter 3—"Data Evaluation"—for more on misinformation.

22. Mark Levene, *An Introduction to Search Engines and Web Navigation*, second edition (Hoboken, NJ: John Wiley & Sons, 2010), 10.

23. Mark Levene, *An Introduction to Search Engines and Web Navigation*, 10.

24. IP address is the abbreviation for Internet Protocol. It is the logical address assigned to a device, like a computer, on a network. Every computer on the Internet has an IP address, and that address is unique. Therefore, unless a counterintelligence strategy is used to block, hide, or disguise this IP address, its discovery by others can lead them to the intelligence analyst making the inquiries.

25. For a detailed discussion about defensive as well as offensive counterintelligence strategies, see Hank Prunckun, *Counterintelligence Theory and Practice* (Lanham, MD: Rowman & Littlefield, 2011).

26. These would, of course, be unclassified data warehouses, not those of law enforcement, the military, or national security agencies. Nevertheless, these agencies would no doubt find some commercial databases very attractive to the types of issues they are probing and hence may have commercial agreements in place to access business-related data sets as any private contractor may.

27. See also the list of other potential sources of information in the section entitled "Other Sources—Where's What" earlier in this chapter.

28. Michael C. Taylor, "Doctrine Corner: Open Source Intelligence Doctrine," *Military Intelligence Professional Bulletin* 31, no. 4 (October–December 2005): 14.

9

๑

Clandestine and Covert Sources of Information

This topic examines secret sources of information in the context of the wider spectrum of data available, including information that can be obtained from:

1. Clandestine and covert defined;
2. Undercover operatives;
3. Informants and agents;
4. Physical surveillance;
5. Optical surveillance;
6. Covert photography;
7. Aerial photography;
8. Electronic surveillance;
9. Audio surveillance limitations;
10. Placement of surveillance devices;
11. Mail covers; and
12. Waste recovery.

CLANDESTINE AND COVERT DEFINED

The spectrum of information sources ranges from open and semi-open sources to clandestine and covert. Having examined open source information in chapter 8, this chapter discusses sources at the other end of the information spectrum that are available to the intelligence analyst—those considered to be surreptitious in some way. Because of the intrinsically safe nature of open source information, the same high level of consideration and planning is required to obtain these data when compared to organizing data collection by clandestine and covert means.

Clandestine data collection, although akin to covert collection, is different in that clandestine operations operate in the open—visible to the target but disguised so that they do not appear to be what they seem. *Covert* operations, in contrast, are carried out in secret. They are hidden and not visible to the target even in a disguised form; they are to some degree intrusive, but because they are invisible, the target has no knowledge that an operation is being conducted.

These methods are discussed here and include undercover operatives, informants/agents, physical surveillance, electronic surveillance, informants, mail covers, and waste collection. Analysts may want to use these methods in their collection plans, because an attempt to obtain information via open methods might be met with a nil result. For instance, secretive methods are used where the target is concealing information. The only way to obtain such data under these circumstances might be to penetrate the security measures via a surreptitious method.

UNDERCOVER OPERATIVES

The objective of the undercover [operative] is to infiltrate "as deep as possible and [gather information] on the opposition or enemy."[1]

1. J. Kirk Barefoot, *Undercover Investigation* (Springfield, IL: Charles C Thomas, 1975), 4.

Undercover operatives are able to get close to individuals or inside organizations to make firsthand observations. The use of operatives is risky for the operative and the organization by which they are employed. These risks are both physical and psychological.

The operative risks physical harm in the form of bodily injury and death as well as a range of psychological injuries spanning from mild anxiety to severe psychiatric disorders. The physical risks are more apparent, as one can easily visualize the ramifications of having to penetrate an illegal enterprise. The psychological injuries arise from the stresses associated with working in isolation, working in a dangerous environment, and perhaps engaging in activity that is illegal (including consuming illicit drugs and alcohol in binge quantities).

The data that can be obtained from an undercover operative can be very valuable because it is an opportunity to get a glimpse of the target's intentions, thereby providing an insight into the target's thinking, rationale, and behaviors that could not be obtained by other means. However, given the risks and the monetary costs of "running" a field operative, it is a method that is not often used in the first instance. It is usually reserved as a means of last resort or for targets that pose a serious danger to society or national security.

Because the operative will be in direct contact with the target, a number of issues must be kept in mind. The most important is that the operative's identity must be guarded with utmost secrecy. If knowledge of the operative leaks to the target, the operative's cover will not only be "blown," but also the operative is likely to suffer injury or death.

Part of the operative's brief is to obtain evidence of wrongdoing in a law enforcement context, or classified intelligence in the case of national security. Evidence may be in the form of admissions, but intelligence data may be in the form of indicators of intent. To capture these data, the operative can either commit the details to memory and then record them later for transmission back to the analyst's agency or use some electronic device that transmits the data live for recording and transcription. The latter is the most reliable and the best solution, as it doesn't rely on the operative's ability to remember the details, which, from an intelligence point of view, can be critical. Data can be qualitative (e.g., discussions with the target) and quantitative (e.g., numbers of items, times, routes, colors, preferences, and so on).

INFORMANTS AND AGENTS

Using informants, or agents, to gather information places a safe distance between the agency's operations and the target. An informant is someone who has indicated willingness to assist the agency in achieving its goals, and is similar to an agent who acts on the agency's behalf (a proxy)—whether that is to arrest a drug trafficker, close down a trademark infringement ring, or sell state secrets to the agency or its government. Informants and agents do these things for a variety of reasons. As T. J. Waters discussed in his memoirs, *Class 11: Inside the CIA's First Post-9/11 Spy Class*:

> Values + beliefs = behavior. This is what you need to keep in mind. Know your target's motivations in the context of their values and beliefs. Values govern behavior as motivation for our actions. Beliefs are how we express

our values to ourselves and others. . . . We must understand the agent's history, the background fundamentals of how they became who they are in order to understand their motivations.[1]

The processes for informant/agent handling vary from agency to agency and from informant to informant, or agent to agent, but in general, the process of employing these people starts with some form of "registration." This serves several purposes, the main being accountability of the funds (or other gratuities) that will ultimately be paid to them for service.

Corruption is a temptation when running informants because the system relies on the honesty of the agent handlers to pass over the full amount of funds without "skimming" any for themselves. Also, there is the temptation to run fictitious agents in order to collect payment; this ruse was made legendary by Graham Greene in his book *Our Man in Havana*, whose main character, Jim Wormold, ran an entire spy network that did not exist but regularly passed on intelligence reports, which were, of course, fictitious.[2]

Although the *Our Man in Havana* situation may sound like the fanciful plot only found in the pages of a spy thriller, it has a basis in real espionage operations, as confirmed by an ex-case officer: "Most seasoned operations officers[3] have had that unsettling experience of going to a 'cold meeting,' only to find out the 'agent' did not know he had been recruited, or simply did not exist."[4] It is also reaffirmed by a former legislative counsel to the director of central intelligence and deputy chief of operations for Europe who stated, "A CIA spy runner in Western Europe in the 1980s apparently duped his colleagues for five years with lengthy fictitious reports from imaginary European agents on local attitudes toward NATO and scepticism about U.S. commitment to come to the defense of its European allies in the event of a Soviet attack."[5]

As such, informants are photographed, fingerprinted, and their vital statistics recorded on an official register. This register is maintained by the agency, and the data it contains are classified so only those that have a need-to-know can access it. This helps to protect the agent's identity and, therefore, safety.

The data that informants can provide will vary greatly—from accurate and reliable down to inaccurate and unreliable. If the agents are not trained in observation skills or memory retention techniques, the data they will provide may be sketchy. If, however, they are trained—as in the case of a foreign intelligence operative who has "swapped sides"—the data may be very sound. But having said that, experience shows that trying to establish the veracity and value of an agent's reports is very difficult. It may also be difficult to establish their bona fides.[6]

To overcome training shortcomings, a listening device may be employed (such as a body transmitter), but this option may be reserved to where the informant is only used once or twice, as the risk of discovery and personal harm is great—with each use, the chance of detection increases. Also, there has to be an agency listening post nearby to receive the transmitter's signals and record the audio. Having a van or other vehicle containing the listening post equipment parked near the target's location may unduly raise suspicion.

If an apartment is rented close by to set up a more permanent listening post, this will incur added financial costs and open up another possible source of leaks—for instance, the realtor or landlord may "talk" if the agency has rented the property openly without using a cover. If the agency has rented the premises under a cover, the foot traffic in and out of the building may expose the operation. As with undercover operatives, the use of an informant or agent to gather data must be weighed against a number of risks: exposure of the operation, harm to the informant, the likelihood of obtaining the information sought, and the likelihood of obtaining that information accurately.

In regard to the last point, it will always be the case that the analyst needs to try to verify the accuracy and reliability of the agent's data. There may be many ways of doing this, but one example is to compare the data against similar information provided by the same source. Whatever way it is done, it should be borne in mind that there is some degree of uncertainty inherent in running informants. Their motivation can be greed, revenge, or exchange for some favor (e.g., reduced jail sentence), anything *but* the pursuit of truth.

PHYSICAL SURVEILLANCE

Operatives "assigned to a surveillance operation should possess a reasonable amount of resourcefulness and adaptability so that [they] can effectively blend with the environment, both in appearance and conduct."[2]

2. Raymond Siljander, *Fundamentals of Physical Surveillance: A Guide for Uniformed and Plainclothes Personnel* (Springfield, IL: Charles C Thomas, 1977), 5.

Physical surveillance is the act of making observations of people, vehicles, or the activities occurring at specific locations. Many research methodologies use observation as the key means of collecting data. This can take the form of a researcher standing on a street corner and counting pedestrians as they pass by or as they enter a building, or it could be a data collector who records the number of vehicles that pass a particular intersection or travel along a suburban street. Surveillance used in intelligence operations is usually covert and takes the form of either moving surveillance or fixed surveillance.

Physical surveillance may take place either at a particular location, which is known as a *stakeout*, or in a continually moving situation, referred to as a *tail*. Physical surveillance is often used to supplement information which has been obtained from open sources. It can also be used early in an operation to accelerate the generation of leads, corroborate existing information, or obtain details that would not be available through other avenues of inquiry.

Fixed surveillance is a tedious and time-consuming activity, but its value should never be underestimated. Its strength lies with the field operative, who is tasked with a single mission—to obtain details about the target. The results can be pivotal to any intelligence research project. Unlike social research where a data collector stands in the open with a clipboard and counts passersby, the covert surveillance officer, termed an *operative*, may sit for hours in the darkened interior of a van or in the backseat of a car, video recording the comings and goings of the target and associates.

Moving surveillance is when the operative follows the target. The objectives of the moving surveillance are the same as a fixed surveillance but, in addition, the target is tailed wherever they go. The most common form of moving surveillance is via car, as this is the usual mode of transportation people use. Surveillance on public transport—aircraft, bus, train, and tram—is also possible, but the level of risk of detection increases, as following a target in such close proximity makes the operative's task more difficult.

A benefit of covert surveillance is to validate the information coming from undercover operatives and informants/agents. For instance, a surveillance operative should be able to independently verify that, say, an informant did meet with the target on the day and time provided by an informant. Although it is highly unlikely that the operative will be in a position to verify what was said between the parties, validating the meeting may go a long way in establishing an informant's creditability, especially if the pattern is repeated. However, there may be other ways of validating the details of a meeting if a surveillance operative is able to video the meeting.

For instance, the length of the meeting may be able to suggest whether the details the informant has purported during the conversation are accurate—a two-minute conversation on a street corner outside a coffee shop is not consistent with details of a long planning meeting over a lengthy meal at a restaurant. Surveillance would be able to shed light on this. Likewise, the body language of the two as they talk may be an indicator of how well the meeting went—was there aggravation, frustration, disbelief, threats, or did it conclude with mutual admiration, gratitude, and a relaxed atmosphere? Such independent observations could prove valuable in supplementing or validating other covertly obtained data. The important point is that surveillance is actual observations, not self-reported information; the latter can be subject to any number of biases.

Vehicles in the Open

A target's motor vehicle is another source of potentially valuable information. Walk down a street in the business district of any city or town and view the interior of the cars parked at the curb as you pass. Chances are that you will see many personal and perhaps confidential items left in plain view. Backseats, front seats, and dashboards can be littered with objects and papers that could provide rich information, or act as leads in finding other sources of information. Permits affixed to windows have numerous facts that are potentially of interest.

This is a form of physical surveillance not often thought about or incorporated into an information collection plan—perhaps because it is so obvious that analysts overlook it as a source of data. Unless the target practices counterintelligence, there are usually a few pieces of information that could be helpful to an intelligence research project. If the vehicle is devoid of objects or papers, this could mean that the target is fastidiously tidy (which might be valuable in itself), or that he is practicing sound operational security. Having said that, if the target is suspected of being one who is cunning (e.g., a foreign intelligence service's operative or a terrorist), and investigators obtain information from a vehicle "in the open," then be aware that it could have been planted as a means of deception (see deliberately deceptive information in chapter 3 under the heading of "Data Evaluation").

OPTICAL SURVEILLANCE

Because operatives rely so heavily on their eyesight in conducting physical surveillance, various devices are used to assist them in extending their vision. Using technology also assists operatives in positioning themselves much further away from the target than could normally be achieved with unaided sight, thus reducing the possibility of discovery (see figure 9.1).

The principal and traditional device used in physical surveillance is a pair of binoculars.[7] Binoculars are an optical device consisting of two prism-operated telescopes fixed in parallel. This configuration enables an operative (sometimes termed the *surveillant*) to view a magnified image of the subject using both eyes. Binoculars have been designed to provide both magnifying power and light-gathering capabilities. The latter is essential for surveillance work at dusk and at night. Binoculars that have a built-in digital camera are also available commercially.

Viewing a subject at distances that exceed the effective range of binoculars is accomplished by using a telescope. The magnifying power of the telescope ranges from twenty to several hundred times that of normal vision. Binoculars, in comparison, range from about six to twenty times that of normal vision.

A viewing device usually at home in submarines, but sometimes used in land-based information gathering, is the periscope. Small, portable, high-quality devices are used to peer into high, inaccessible windows, over walls, and around corners. Periscopes are also used in applications such as surveillance vans (e.g., mounted on the roof). With the availability of so-called "action cameras," a form of periscope can be constructed by placing the camera on the end of a pole (which can be telescopic for concealment in transport). The camera can then be lifted over the obstructing wall, fence, etc., or lifted to view through a window that would be too high to see into otherwise, and the target area recorded for analysis.

Continued developments in optical technology have now made low-light (night) viewers available at affordable prices. These units are designated as either "active" or "passive" night vision devices. The first group, the active devices, are devices that operate by using an infrared light beam. The surveillant projects the invisible beam of energy so that it illuminates the subject. The image is then viewed with special equipment, which converts the infrared radiation into the visible light spectrum. The second range of devices operates by amplifying the existing background light—moon, stars, streetlights, and so on—by several thousand times, thus literally turning night into day through the sights of the surveillant's night scope.

COVERT PHOTOGRAPHY

In the context of intelligence work, covert photography refers to what is generally termed *surveillance photography*.[8] In conducting covert photography, by far the most commonly used camera is a digital single lens reflex (SLR) (which replaced the 35-millimeter-film-based camera). However, in some unique situations, such as surreptitious copying of documents, a

Figure 9.1. Covert optical surveillance of illegal fishing. Port Roper near the Gulf of Carpentaria, Australia.

Photograph courtesy of Intelligence Officer Nathan McGrath of the Northern Territory Police, Fire, and Emergency Services, Australia

subminiature camera would usually be employed. The value of subminiature cameras lies in their concealable size or concealable form (e.g., mobile/cell phones with built-in cameras). Nevertheless, the advantage of digital SLRs is their fast (light-sensitive) lenses. The ability to download images, edit them using commercial software, and send them over the Internet applies to all digital photography.

As with binoculars and telescopes in physical observation, telephoto lenses play an important role in covert photography. A reflective mirror lens (catadioptric) not only enables a surveillant to greatly reduce the physical size of his equipment, making concealment easier, it acts to extend the operational distances. Basically, catadioptric lenses use a system of mirrors to compress the light's optical path. Magnification (measured in the focal length of the lens) varies from about 100 to 2,000 millimeters.

Digital video cameras are also popular for surveillance work. These cameras are no larger than SLRs—and, in a growing number of models, smaller—so they are often used with a telephoto lens, thus allowing surveillants to record their observations from safe distances. As digital technologies continue to develop, it is likely we will see cameras with higher-definition resolutions and they will be more affordable. The application of these devices for covert surveillance will also increase the span of applications.

Many mobile/cellular telephones have video recording capabilities as part of their functioning and therefore make ideal covert cameras (though there is a degrading of image quality when compared to a dedicated video camera). Eyeglass-style digital recording devices are being developed that can take photographs and video, as well as access the Internet. Information and images taken by these devices can be uploaded to and downloaded from the Internet via the wearer's voice commands.

In the case of a fixed position surveillance (a stakeout), a time-lapse option can be used to provide extended coverage from a single digital memory device. In order to achieve this, time-lapse photography operates in a frame-by-frame mode that greatly reduces the demand for electronic memory storage (twenty-five frames per second is considered real-time recording). Many state-of-the-art video cameras can be built into such items as briefcases, books, wall clocks, smoke alarms, writing pens, paintings, and plants.

A growing number of devices are also marketed as "action cameras" that can be attached to parts of vehicles—e.g., windows—as well as personal equipment such as helmets, utility belts, and so on. Take for example the case where a surveillance operative needs to watch a fixed location for long periods of time, he or she would normally be required to sit in their car or van and use various screening devices, like window shades, so they remain hidden inside. But they also need to be able to immediately take photographs of people or events. However, with an "action camera" mounted on, say, the dashboard of a car, the vehicle could be parked overlooking the target area and the camera angled to record the desired field of view (see figure 9.2). Many hours 'or perhaps days' worth of digital images could be obtained while the car is parked unoccupied, and once retrieved, could be fast-forwarded to the event(s) of interest. Variations of this technique are possible, thus reducing suspicion and hence the chance the operative is discovered, which would result in exposing the operation.

AERIAL PHOTOGRAPHY

Aerial photography from both rotary and fixed wing aircraft is another effective method of information gathering. The history of aerial surveillance dates back to the mid-nineteenth century when the first photograph was taken from a French military (hot air) balloon.

Since that time, many developments have occurred in the area of photoreconnaissance. The technology currently available ranges from

Figure 9.2. Traditional covert photography. Here an operative is positioned in a car observing the target. This situation could be replaced by an unattended vehicle with an "action camera" mounted on the dashboard.

Photograph by the author and provided by courtesy of CC Thomas Publisher, Ltd

reconnaissance satellites and ultrasophisticated spy planes down to the light plane-for-hire with conventional handheld photographic equipment. The former, of course, are used by intelligence agencies of various nations, while the latter would probably be used on a rental basis by smaller agencies on an *ad hoc* basis.

At one time remote-control model aircraft were the province of model enthusiasts. However, a number of commercially available remote-controlled aircraft of varying sizes are equipped with digital cameras for carrying out surveillance work. There are both rotary wing and fixed wing versions, and both varieties function in the same way as military unmanned aerial vehicles (UAVs), or as they are commonly referred to, *drones*. These drones are capable of carrying multiple cameras, individually controlled so that several ground targets can be tracked simultaneously. Software then can be used so that each camera maintains focus on a particular target of interest; or it can be programmed to have a camera, or cameras, scan for targets that meet a preset set of conditions. Once these visual data are obtained, other software can be used to generate, for instance, social network charts of the people observed.

Internet-based mapping facilities give analysts an easy and cost-effective option for obtaining aerial photographs. Websites provide satellite images of the earth and cover almost all locations on the globe. Intelligence analysts can select several different types of views: satellite imagery, maps, and terrain, or combinations. Street-level views are also available. Although these sources would not be practical for the analyst who is monitoring an arms control agreement (for the most part, these are static images refreshed only occasionally), they may be all that is needed to brief a local police or private investigation surveillance team as to, say, reconnoiter the terrain surrounding the target's place of business. Some of these are of very high quality, so much so that the news media reported that Norway's National Security Authority blocked U.S. company Apple Inc. from flying over Oslo—the nation's capital—to take three-dimensional photographs for use in its mapping application. It was reported that the intelligence agency feared these high-quality maps could be used to undermine national security.[9]

ELECTRONIC SURVEILLANCE

As a general principle, where the opposition has established a high degree of operational security (i.e., defensive counterintelligence measures), the penetration of human operatives and agents will be prevented. This may also be a result of other factors, such as cultural, ethnic, religious, language, or other natural barriers. In these situations, data collection methods often shift to more technical means. But this shift should not be at the total expense of attempting to have human sources in place who can provide reports based on direct observations. The relationship between operational security and a shift in technical collection methods is shown in figure 9.3.

Electronic surveillance is most likely to feature prominently in any shift to technical collection. This may be in the form of audio surveillance (i.e., room listening devices or *bugs*), telephonic intercepts, or Internet-based intercepts. There are other forms of electronic surveillance, such as radio frequency intercepts (i.e., the interception of radio transmissions), but the most commonly used intercepts are from these three methods.

Audio Surveillance

The fundamental principle of any audio surveillance operation is to be able to plant a quality microphone as near as possible to the target and, in doing so, avoid background noise that may render the intercept unintelligible. The types of microphones used in audio surveillance are

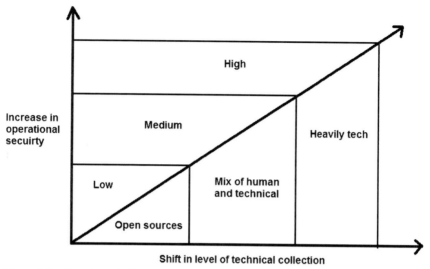

Figure 9.3. Relationship between operational security and technical data collection methods.

required by the nature of the work to be very small and are referred to in the trade as *subminiature*. Their minute size allows them to be secretly deployed in the target's environment. Once in place, they can be connected to a high gain amplifier, then to headphones (for live monitoring), a digital recorder (for listening at a later time), or a transmitter to relay the audio to a listening post.

If a transmitter is not used to send the intercepted audio, then wires are needed; the wires leading from the microphone to the listening/amplifying equipment must be concealed to prevent detection. In some cases, existing wires such as telephone, electrical, or even, paradoxically, burglar alarm wires can be used instead of running new wires, thus making detection more difficult.

There are also special metal paints on the market which an operative can use to "paint" wires across a room. Once touched up with paint of the surrounding decor, these "wires" are reported to be very difficult to detect (especially if used in an industrial or warehouse setting). There are several types of microphones the surveillant can use, depending on the particular situation. They are the tube, contact/spike, pneumatic, and directional microphones.

Tube microphones are designed to be inserted into a targeted room via a very small drill hole. This species of microphone consists of an element connected to a thin tube. The tube emerges flush with the wall in the targeted room and can only be detected by very close inspection. Tube microphones are likely to be located in spots made classic in spy novels—behind a wall, a picture, a piece of furniture, anything that would hinder detection for as long as possible.

Contact and spike microphones, in contrast, require less effort to secure their installation. These microphones do not respond to air vibration as a conventional microphone does but rather translate vibration into sound. Contact microphones are similar to the "pickups" used by musicians to amplify their instruments, and spike microphones resemble the vintage phonograph needle. These types of microphones are attached to the exterior of a wall, window, floor, or ceiling. Once in place, these devices will reproduce quite clearly the sounds produced within the targeted room. These microphones lend themselves well to permanent installation.

The pneumatic cavity microphone is the electronic version of the drinking glass placed against the wall trick, historically recognized as an effective method for monitoring adjacent room conversations (figure 9.4). The pneumatic microphone is substantially superior and operates by using a specially constructed shell which is highly responsive to surface vibrations at audio frequencies found in the range of human speech. This "cavity" is used in conjunction with a conventional microphone element to enhance the device's performance. It also forces audio output to correspond to a wall's surface (or window, floor, or ceiling, depending on the case) vibrations rather than a direct sound output.

Directional microphones are of two types: parabolic and shotgun. Parabolic microphones consist of a "dish" with an inwardly pointing microphone element. The targeted audio is reflected and focused into the microphone element, thus gaining a directional effect. The shotgun microphone (also known as a *rifle* or *machine gun* mic) operates on the same principle but utilizes a long tube, or set of tubes, in a cluster to pinpoint the targeted conversation. The effective range of these microphones is reported to be about 150 meters, and they are known to be able to pick up audio through closed windows at closer distances.

Radio transmitters offer an operative a much greater degree of safety from detection because installation requires far less effort than a hardwired device.

Another area of audio surveillance, discussed briefly above, is that of wireless microphones, also known as miniature radio transmitters. These devices rank further up the hierarchy in surveillance sophistication. These devices do not have any wires or connections to reveal their location.

Figure 9.4. "The walls have ears." A simple but effective method in a high-tech world—the drinking glass placed against the wall trick.

Photograph by author

They can be attached to furniture or fixtures by means of a magnet or adhesive surface or concealed in everyday objects such as rings, pens, cigarette lighters, books, ashtrays, or pictures. Transmitters do not require the eavesdropper, or his equipment, to be located nearby. Their range of transmission varies from sixty meters to almost a kilometer. It is directly dependent upon transmitter strength, the thickness of surrounding walls, the sensitivity and selectivity of the receiver, as well as its antenna system.

Body transmitters are generally larger, more powerful, and better constructed than wireless microphones. This is because the chance of detection is slight, and the device is intended to be used repeatedly. These devices are designed to operate from an operative's coat pocket, underclothing, or attached by tape directly to the operative's body.

Akin to the body transmitter is the briefcase transmitter. Not only can this device be used by a "walk-in spy," but it can be conveniently "forgotten" in the target's room or office in order to obtain ensuing conversations. Another type of electronic surveillance transmitter operates by broadcasting in the very low frequency range (VLF is between 3 kHz and 30 kHz). This device uses electrical power lines for signal transmission. The signals move along the wire path, and, because of the device's very low frequency, very little energy is radiated into space. This method of communication is used by many of the household wireless intercoms sold commercially.

Communication equipment which operates outside of the standard FM radio broadcast frequencies tends to be more secure from interception (that is, outside of the 88 MHz to 108 MHz frequency range). This is because the radio receivers needed for this type of radio reception are not sold in regular retail outlets. Nevertheless, they are available commercially at radio and electronics stores and are a common feature in the radio rooms of radio amateurs (i.e., "ham radio").

There is also a cadre of radio enthusiasts in the general community known as "listeners" who make it a hobby to scan the radio spectrum listening for any unusual content or sources (e.g., a shortwave listener). These hobbyists have high-gain antennas and wideband receivers that scan many hundreds of individual radio frequencies each second (commonly termed *scanners*) and, as such, could detect the conversation collected by a bug (see figure 9.5).

If there is a chance of interception by a casual listener, the radio technician who will oversee the bug installation should consider the use of a scrambler or other form of encryption. Contrast this countermeasure with the failed Watergate political espionage operation of 1972, where the operatives used no operational security to protect their two-way radio conversations.[10,11]

Some receivers operate in the microwave part of the radio spectrum, such as those used by telephone companies for telecommunications and by private enterprises for computer data transmission. Such receivers are complicated in design and expensive, although military and governmental intelligence agencies would have ready access to these. Units used for intercepting microwave communications can be set up in a van or building anywhere along the path between the transmitter and the receiver, which could be several hundred kilometers long.

Several devices, although not specifically designed for surveillance work, are worthy of a brief note because their function provides the operatives with an increased scope of application. These are the drop-out relay, the voice-operated relay (VOX), and the carrier switch.

Figure 9.5.   Icom IC-R72 wideband communications receiver capable of AM/N, FM, SSB, and CW/N reception from 30kHz to 30MHz
Photograph by author

The drop-out relay is attached to the target's telephone line and then wired to a transmitter. It will switch the bug on whenever the hand piece is lifted from its cradle and off when it is replaced. This prolongs battery life, and because the bug is not transmitting continuously, it lowers the risk of being detected by an electronic countermeasures sweep.

A voice activated switch (VOX) is similar in purpose to the drop-out relay. If connected to a room transmitter, the bug remains dormant until activated by the sound of a voice or noise. The VOX turns the bug on and off as individuals enter and leave the targeted room.

Finally, a carrier switch can be used to start and stop a digital recorder when it receives an audio signal, and it can be activated by a hidden transmitter. Employed in conjunction with a drop-out relay or a VOX, the switch carrier increases exponentially the surveillance capabilities of the operative.

Telephone Intercepts

The telephone is a very useful medium for electronic surveillance. Obtaining information using the telephone involves two methods: the first uses devices that intercept conversation directly from landlines and requires no entry into the target's property, and the second is one that uses a portion of the telephone system for room eavesdropping and usually requires physical access to some part of the telephone system.

Wiretaps

Wiretapping is the interception of fixed line telephone and facsimile communication, as well as computer data that are transmitted over landlines. The interception of these signals can take place anywhere between the target's location and that of the surveillant. The more difficult parts of the telephone system to install a device are at the target's property and the lines leading out of that building. Once the targeted line(s) leaves the building and melds with the wider telephone network, interception is far less difficult, because with a court-ordered warrant, a surveillant can install unobtrusive equipment with greater ease.

Because the telephone company provides all of the electrical power required to operate a target's telephone service, this medium offers great eavesdropping benefits to operatives. For example, the telephone's power supply can be used to directly operate electronic eavesdropping devices, the wires themselves can be used to carry the resulting audio signals, and the microphone in the handset can be used to listen in on room conversations.

The techniques used to tap conversation from telephone lines consist of (a) direct wire connections and (b) induction coils. In direct wire connections, the lines are cut and the listening device spliced in place, using an electronic matching network. With induction coils, the tapping process literally lifts the audio signals off the telephone line and, therefore, does not require splicing. For this reason, induction coils can prove to be undetectable by electronic countermeasure sweeps. The only effective way to detect this type of bug is by visual inspection of the telephone wiring.

An alternative to direct wire connections is a radio system. This is the same as the wireless microphones described previously, except that it requires no microphone element. This is because the audio signals are already in an electrical form. Radio systems use the telephone line voltage for power, as opposed to batteries, and transmit whatever conversations are on the targeted line to a remote receiver/digital recorder.

The placement of telephone surveillance devices can be quite arbitrary; the only limit to their deployment would be the ability of an operative to gain access to the target's telephone system. The devices may be installed within the telephone instrument itself, anywhere along the line in the targeted building, on a telephone pole outside the building, or in the wire closet or terminal room where the lines are joined to the branch feeder cable.

There are several other electronic telephone surveillance devices which allow an operative to record the telephone numbers and dates when dialed by a target but do not permit the recording of the conversation. These are the dial impulse recorder, commonly referred to as a pen register, and the touch tone decoder. They operate simply by counting the impulses in each dial pulse group, that is, the digit dialed. An alternative to these two devices is the variable speed tape recorder. This operates by recording the desired conversation, then replaying it at a reduced speed so that the impulses can be counted to determine the number dialed.

The harmonica bug, or infinity transmitter, is an eavesdropping device used for planting within the telephone instrument itself. This device consists of a tone-controlled switch, coupled with an audio amplifier and a microphone. Although it uses the telephone system, the device functions as a room eavesdropper as opposed to a telephone conversation interceptor. The infinity transmitter uses the existing telephone lines for conveying the surreptitiously acquired conversation. It is activated from an infinite distance by a tone generator similar to those used by answering machines. The tone is said to be a note produced by a harmonica, hence its alternative name.

In order to operate the infinity transmitter, the operative dials the target's telephone number, which can be local, interstate, or international. After dialing the number, but before the telephone rings, a tone is sounded into the operative's telephone mouthpiece. On the target end of the telephone, the infinity transmitter receives the audio tone and switches the device to answer this telephone electrically rather than physically. If this is performed correctly, the target telephone should not ring. This, in effect, means that the telephone is working even though the hand piece still remains on the cradle. Once operational, the surveillant may monitor the room conversation. If the subject attempts to use the telephone, the operative simply hangs up, and the device is electrically disconnected, returning the instrument to normal operation.

There are several other techniques directed at modifying the telephone instrument for eavesdropping. This form of electronic surveillance exploits the normal operation of the entire telephone system. By shorting or bypassing the hook switch on the instrument, the telephone becomes a live microphone. This technique is known as telephone "compromising."

Another dimension to electronic surveillance is systems that operate by using directional beams of light energy. These systems are based on the use of laser beams of either visible or infrared energy to convey their intercepted audio. They are reported to be quite reliable and virtually undetectable. The system consists of a laser light source, which focuses its beam on a window in a targeted room, and an optical receiving/decoding device. The system operates by detecting minute vibrations of the reflected beam caused by the room's audio. Decoding of these vibrations reproduces the conversation or sounds within.

Interception of computer data can be accomplished in similar ways to eavesdropping on room conversations. With data interception, the legitimate user's data transfer is intercepted through a suitable wiretap or bug and either recorded or transmitted to a listening post as is done in the case of audio. With the interception of computer data, an operative monitors the data exchange from terminal to main computer, thereby tabulating input data and computer responses.

Pen Registers

There are also devices that allow operatives to collect data relating to the telephone number dialed by a target (known as a *pen register*) or to collect the telephone number that dials the target (known as a *trap-and-trace* device). The possession and use of these devices are governed by legislation, and laws vary from jurisdiction to jurisdiction.

Pen registers are usually manufactured to only record details such as the target's telephone area code, number, and, where an extension telephone is used, the number of the extension dialed. In the main, the reason why pen registers are not capable of collecting what is termed "transactional data" is that the data input by touchtone telephones—for instance an account number or a PIN—would require another form of court order (i.e., warrant). Therefore, a pen register that is only capable of recording numbers avoids inadvertent breaches of an electronic interception warrant.

Internet-Based Intercepts

Apart from some university and government research libraries, the Internet is arguably the wealthiest source of information for analysts. It is not only a source for published material but also the private messages transmitted over the Internet (e.g., e-mail). Businesses, governments, and individuals all create, access, or modify records of numerous descriptions that are located on computer servers connected to the Internet. Because these data exist in digital form on magnetic media, this realm is known as *cyberspace*.

In general, analysts can obtain data without a warrant if the data are placed in the public domain. This is analogous to the physical world where access is not restricted—public libraries, commercially published books and magazines, and newspapers. However, where access could be considered "private" by a court, then some form of legal order or subpoena is likely to be needed. By and large, the laws pertaining to telephonic intercepts apply to intercepting data in cyberspace; in many jurisdictions, laws have been enacted to cover such intercepts. Analysts must always work within the law when formulating information collection plans. When in doubt, obtain a legal opinion as to the legality of what is being proposed, especially if it is a new, novel, or previously untested method of data collection.

AUDIO SURVEILLANCE LIMITATIONS

For obvious reasons, it is impossible to determine the extent of electronic eavesdropping that is carried out by the intelligence community each year, although from media reports, it appears to be quite widely used and not limited to any one intelligence type. If a target suspects he or she is the subject of a wiretap or bug, an audio countermeasure sweep is the usual way the target will try to determine if an operation is under way.

The sweep, however, will reveal devices operating at that given time only. It must be stressed that no room can be guaranteed to be proof against audio surveillance. In the past, even the most sensitive rooms in the U.S. embassy in Moscow have been reported to have been penetrated. Conducting sweeps at nonconstant intervals is the target's most effective way of countering an intelligence agency's audio surveillance. It is the most reliable way to check for, and clear, audio surveillance devices.

There are limitations, however, to this type of countermeasure. Firstly, with regard to telephones, even if the target's telephone instrument(s) appears to be clear at the time of a sweep, there is no way of determining whether the telephone of another party is under surveillance by inspecting the target's end of the line. It is understood that there is also no technology available to date which can check for listening devices at, or beyond, the central telephone exchange. Secondly, there are some state-of-the-art devices and techniques used by intelligence agencies that may be undetectable because of their high level of technical sophistication.

Audio countermeasure sweeps are conducted by both specialist business counterintelligence firms and by private investigators. Such services are usually listed in the Yellow Pages of the telephone directory. A professional sweep usually includes both a thorough physical search—inspecting literally every inch of and every object in the suspected area—and an electronic sweep. The electronic sweep may utilize a broadband receiver (like those used by ham radio operators) and/or a specially designed field-strength meter to test for transmitters. Metal detectors can be used to hunt for bugs in nonmetallic objects and deeply planted devices in walls, floors, and ceilings. There are also a wide range of diagnostic meters used to test the telephone line voltage for the presence of wiretaps.

PLACEMENT OF SURVEILLANCE DEVICES

If audio surveillance devices were to be placed in the target's property, it could be done by one or more of the following time-proven espionage methods:

- Friendly access;
- Surreptitious entry;

- Infiltration; or
- Secreted in a gift.

Friendly Access

Access to a target's building may be limited to employees and visitors who are known or have appointments. All other visitors may be screened and their identities verified prior to entry. People making deliveries, including mail deliveries and maintenance workers, may be handled in the same manner. Access to offices could be on a restricted, need-to-be-there basis. If visitor/staff traffic is heavy, a system of custom-designed identity cards to be worn by employees would be an efficient method of establishing "friend" or "foe." Toilets and other isolated places might be checked at the end of the day's business for operatives hiding in the building.

Surreptitious Entry

Break-ins and burglaries are not an uncommon occurrence for businesses and private homes to experience. But in June 1972 Watergate underscored the reality that break-ins are not only a method for acquiring cash and valuable physical assets by common criminals, but also are a technique for intelligence gathering.[12]

However, as the former director of central intelligence, the late Richard Helms, once pointed out, these are technically difficult operations where the risks must be balanced with the anticipated gain. Photographing files, bugging telephones, and installing electronic monitoring equipment takes planning and the resources of a well-financed and well-equipped agency. The agency also needs well-trained operatives who are supported by competent technical engineers.

In intelligence work, this technique is referred to as a *black bag operation*. Surreptitious entries are used to plant surveillance devices or to carry out other covert intelligence-gathering activities. From the viewpoint of the operative, this is encouraging, as short of creating a mini-fortress, there is nothing that will make an office 100 percent burglar-proof—even Buckingham Palace has had its intruder.

Infiltration Using a Pretext, Ruse, or Disguise

A *pretext* offers an operative acting on behalf of the intelligence analyst a plausible, commonsense technique for obtaining confidential information. A pretext is any act of deception—ruse, subterfuge, ploy, trick, or disguise—that allows an operative to solicit information by a false reason.

This includes entering premises for obtaining information or being in a place that an operative wouldn't otherwise have access to or permission for.

> "The art of using pretexts is a science and should be approached as one."[3]
>
> 3. Greg Hauser, *Pretext Manual* (Austin, TX: Thomas Investigative, 1994), 5.

Pretext should not be confused with the term *social engineering*, which has gained popularity in recent years. Social engineering is a slang term that commonly refers to an individual act of manipulation (usually for fraudulent purposes) to gain unauthorized access to IT systems. This is vastly different from its true meaning, which is large-scale societal planning. So the use of the term *social engineering* in the context of accessing information surreptitiously is incorrect. The technique is nothing more than a ruse, subterfuge, or pretext. In fact, *pretext* is the term most used by private investigators—PIs rely heavily on this technique as a means of gaining information about their targets.[13,14]

By Telephone

This method is used by operatives, usually on a one-time basis, in order to obtain general information about a business. It is the safest and most innocuous type of infiltration to perpetrate. This type of infiltration is carried out by simply telephoning the target, then using a pretext, attempting to extract as much information as possible. Several calls could be made over a period of time.

On the surface, individual calls would appear to be unrelated, but each is designed to obtain specific pieces of information. Depending on the pretext and the number of pretext calls made, the depth of information an operative could gather might be limited to general information. The target might have acute security awareness, especially about unknown persons. If the target is suspicious, he or she may try to identify a telephone caller by requesting the caller's telephone number, then verifying it by using an online telephone directory before calling the operative back (known as *confirmation by callback*).

By Mail and E-mail

This is another low-grade form of infiltration. Again, using a pretext, the operative will write to a target requesting information. Security-aware targets will look for the warning signs of a mail infiltration, such as the use of post office boxes, business name "fronts," and out-of-state addresses. As for e-mail, free Web-based e-mail accounts can raise the target's suspicions because these are nonverifiable accounts.

In Person

Direct personal infiltration of the target may follow pretext contacts by telephone and mail/e-mail infiltration and physical surveillance. In this way, an operative can gather enough information to establish a credible cover for a direct penetration, or acquaint them with the information needed to recruit an agent (i.e., a proxy) to carry out the task.

> "The most popular technique for securing information is socializing with competitors in non-business settings. Business people generally view their competitors negatively, believing that they go to much further lengths than does their own corporation in gathering competitive intelligence."[4]
>
> 4. William Cohen and Helena Czepiec, "The Role of Ethics in Gathering Corporate Intelligence," *Journal of Business Ethics* 7, no. 3 (1988): 199–203.

Indirectly

This infiltration technique is complex to organize and run but can yield impressive results. Basically, an operative creates a clandestine business or organization designed to draw the target, the target's business, or a member of the target's staff. The bogus business is controlled by the operative. These clandestine enterprises can be as simple as a trade newsletter or as elaborate as a fully operational business.

Once established, the operative uses this cover to gather the information identified in the analyst's information collection plan. An example of this is the advertising of positions in a new and very attractive-sounding business. The business may offer a salary and fringe benefits package in excess of those offered in the market in order to entice the target. Once the target's curriculum vitae is received, it is analyzed for the desired information. If it does not disclose the information sought, additional information will be requested from the applicant and/or a personal interview conducted. The operative, or someone from this cover organization, would then "pump" the target for information.

Gifts

If a listening device is concealed in a gift that will be presented to the target (a form of a Trojan horse), it needs to be done expertly. There have been numerous cases where gifts have housed listening devices; the most notable was the 1952 presentation of the Great Seal of the United States to the American embassy in Moscow by the Russian government. A target that is security aware is likely to examine gifts well.

MAIL COVERS

Postal

A very useful data collection method is the *postal mail cover*. This investigative technique has been used by law enforcement agencies for many decades. It collects information by recording what is printed on the outer covering of an envelope or package via some form of photographic technique—photocopying, digital scanning, or digital photography. For the analyst, it is important to note that a mail cover operation does not involve reading or recording the contents of the postal item (e.g., letter or card); only the data on the outside are recorded. Nonetheless, this simple and effective method of data collection can reap a wealth of information—return address, postmark-related information (date and place of posting), and any description of the contents (e.g., "do not bend—photos enclosed," card, etc.). Because the mail is not opened, no search is being conducted, only a form of physical surveillance. The targets will never know that their mail is being monitored, so there is usually no administrative disclosure requirement by the agency that it is conducting surveillance. Postal mail covers are usually bound by the laws or regulations of a nation's postal service, and these governance arrangements dictate how and when this technique can be used as well as how long the collection operation is allowed to proceed.

E-mail

In 2013 the American signals intelligence intercept program known as PRISM was exposed. It was alleged that e-mail and other forms of "live communications," such as photographs, video chat, and related formats, were being intercepted by intelligence services since 2007. The arguments relating to the legality of this aside, this type of covert information-gathering operation is a good case in point about the value of the information contained in the metadata of such messages. It is understood that these intercepts were not "reading" the content of the messages, but were reading the data contained in the electronic "envelope" that routed the message, as would be done in a postal mail cover (see section above).

These metadata are very useful if combined with other data sources (see the chapter on open source information—data mining). With as little as one, two, or three pieces of additional information, these metadata can produce very focused leads, or provide insights regarding the target or their activity. The strength of this type of operation is that the interception can be automated so millions of intercepts can be handled daily. Moreover, these metadata can then be easily queried in a data warehouse environment when they are combined with other pieces of information.

WASTE RECOVERY

This is a long-standing law enforcement and private investigator technique that is popularly referred to as *dumpster diving*.[15,16,17] Despite the initial aversion to the thought of rummaging around in someone's waste material, this is a potentially rich and valuable source of information (see figure 9.6). Confidential material of all types can be found in waste—manuals, notes, letters, memos, reports, files, photographs, passwords, identity cards, receipts, schedules, itineraries, telephone numbers, and much more (including computer hard disk drives, USB flash drives, and a variety of data that have been backed up onto CDs/DVDs). The reason for this is that most people believe that once a piece of paper (or an old computer drive) is placed in a waste bin, it has "disappeared." They believe that no one would bother "getting dirty" searching through someone else's garbage.

Recovery can take place at any point between where the waste leaves the target's premises to, and including, the landfill site. If the recovery is to take place on the target's premises, be conscious that there may be legal issues associated with the operation, as a court could find that the material recovered was still in the possession of the target, and hence, a warrant of some kind was required.[18]

Information obtained via waste recovery was at one time considered high-value/low-cost because it yielded more benefit than what it cost to gather it. However, with its popularization in the press and cinema, waste recovery has become more difficult. Government agencies, businesses, and individuals regularly use document shredders and are more conscious of how and what they dispose. Security surrounding waste has improved with commercial-scale confidential document destruction becoming a service that is widely available.

Figure 9.6.   Unsecured dumpsters like this one make for potentially rich sources of information.

Photograph by author

KEY WORDS AND PHRASES

The key words and phrases associated with this chapter are listed below. Demonstrate your understanding of each by writing a short definition or explanation in one or two sentences.

- Aerial photography;
- Audio surveillance;
- Black bag operation;
- Bugs;
- Clandestine sources;
- Confirmation by callback;
- Covert photography;
- Covert sources;
- Dumpster diving;
- Electronic surveillance;

- Fixed surveillance;
- Friendly access;
- Infiltration;
- Informants (or agents);
- Internet-based intercepts;
- Moving surveillance;
- Open source information;
- Operative;
- Optical surveillance;
- Paper trail;
- Physical surveillance;
- Postal mail cover;
- Pretext;
- Radio transmitters;
- Stakeout;
- Surreptitious entry;
- Surveillant;
- Tail;
- Telephone intercepts;
- Undercover operatives;
- Waste recovery;
- Wideband receivers; and
- Wiretaps.

STUDY QUESTIONS

1. Explain the difference between obtaining information by covert methods as opposed to clandestine methods.
2. Explain several different types of covert data collection available to the analyst.
3. Describe some of the inherent limitations of covert data and the security precautions a target might employ to guard against penetration.
4. What are the advantages of covert data, and how could an analyst use these in practice? Give examples.

LEARNING ACTIVITY

Select a public building in your jurisdiction. Use an Internet-based aerial photography facility to create an electronic slide briefing for a notional surveillance team. Provide in your briefing details on access to and from the building, the surrounding terrain, and potential infiltration and exfiltration points. Note any limitations the maps may have as a way of better understanding what can be done with these types of sources.

NOTES

1. T. J. Waters, *Class 11: Inside the CIA's First Post-9/11 Spy Class* (New York: Dutton, 2006), 118–19.

2. Graham Greene, *Our Man in Havana* (London: Heinemann, 1958).

3. *Operations officer* is the current term used for the traditional title of *case officer*.

4. Melissa Boyle Mahle, *Denial and Deception: An Insider's View of the CIA from Iran-Contra to 9/11* (New York: Nation Books, 2004), 231.

5. Frederick P. Hitz, *The Great Game: The Myth and Reality of Espionage* (New York: Alfred A. Knopf, 2004), 124.

6. Richard Helms, *A Look Over My Shoulder: A Life in the Central Intelligence Agency* (New York: Random House, 2003), 33.

7. Monoculars are also popular, as they are smaller and hence more concealable (about half the physical size of binoculars). They are reported to be less obvious when held to the eye for viewing, as it requires only one hand to do this, and in doing so, only one eye is covered, thus making the act of viewing less noticeable to people around the surveillant.

8. Raymond Siljander, *Clandestine Photography: Basic to Advanced Daytime and Nighttime Manual Surveillance Photography Techniques* (Springfield, IL: Charles C Thomas, 2012), xi.

9. "Norway Blocks Apple Aerial Photos," *Sky News*, http://www.skynews .com.au/tech/article.aspx?id=896524, (accessed August 18, 2013).

10. Although there is no evidence that their transmissions were intercepted, these operatives took a great risk in making this assumption. G. Gordon Liddy, *Will: The Autobiography of G. Gordon Liddy* (London: Severn House, 1980), 244.

11. For the break-in of Dr. Daniel Ellsberg's psychiatrist's office in California in September 1971, prior to the Watergate black bag operation (Daniel Ellsberg, *Secrets: A Memoir of Vietnam and the Pentagon Papers* [New York: Viking, 2002]), the operatives used four handheld transceivers sold by the Radio Shack electronics retailer—5-watt, 6-channel TRC-100B that operated on the 27 MHz Citizen Band Service (CB). The frequency the operatives used was shared by the local taxi company (Liddy, *Will: The Autobiography of G. Gordon Liddy*, 165).

12. Helms, *A Look Over My Shoulder*, 8, but see also pages 3–13 for a discussion about the events surrounding this operation-gone-wrong.

13. M. Harry, *The Muckraker's Manual: How to Do Your Own Investigative Reporting* (Mason, MI: Loompanics Unlimited, 1980), 73–78.

14. Hauser, *Pretext Manual*.

15. The term *dumpster* is a genericized name for the trademark Dumpster™. In England it is referred to as *skipping*, as these mobile garbage bins are known there as "skips."

16. John Hoffman, *The Art and Science of Dumpster Diving* (Boulder, CO: Paladin Press, 1993).

17. John Hoffman, *Dumpster Diving: The Advanced Course: How to Turn Other People's Trash into Money, Publicity, and Power* (Boulder, CO: Paladin Press, 2002).

18. Rick Sarre and Tim Prenzler, *The Law of Private Security in Australia*, second edition (Pyrmont, Australia: Thomson Lawbook, 2009).

10

⑤

Content Analysis of Qualitative Data

This topic examines one of the most commonly used techniques for analyzing unstructured data in intelligence research by looking at:

1. Benefits of the method;
2. Psychoanalytic profiling;
3. Thematic analysis;
4. Thematic metrics; and
5. Secondary analysis of data.

BENEFITS OF THE METHOD

Content analysis is the analysis of text contained in documents. Because these data are usually in narrative form (e.g., transcribed oral speeches, media interviews, or open letters to a public audience), the most common form of analysis is via qualitative methods, but analysis can also incorporate quantitative methods or both. The central purpose of content analysis is to develop an understanding of what is contained in the text beyond the superficial message. Some commonly used techniques for content analysis include thematic analysis, indexing, and qualitative descriptive analysis.

PSYCHOANALYTIC PROFILING

More specialized techniques, such as psychohistorical and psycholinguistic analysis, can also be used to provide clues about the target's probable course of action. These techniques are based on an examination of behavior patterns exhibited by the target (which can include a nation, state, or other actor) in previous written (or transcribed oral) communications. The assumption is that behavior patterns are manifested in these communication forms, and through a psychoanalytic profile of the target, the analyst can distill a set of behavioral indicators.[1] These profiles provide insight for basing recommendations for action. This technique requires the analyst to have expert subject knowledge (e.g., clinical psychology or psychiatry) to credibly assess the data. If not, help can be enlisted from private practitioners and university lecturers, depending on the security classification of the project.

> On October 6, 1973, Israel suffered a surprise attack launched by Egypt and Syria—the Yom Kippur War. Israel failed to recognize or misinterpreted indicators that could have provided the insight it needed to stand up troops in preparation for war. If a psycholinguistic analysis had been carried out of the provocative speeches by President Sadat weeks before, it may have revealed such indicators.

Similarly, psychohistorical analysis is based on the assumption that a country's behavior may be influenced by a range of factors that make up its cultural identity, including social, anthropological, political, and noteworthy historical events. For example, a psychohistorical analysis of Latin America would need to consider the issue of U.S.–Latin American relations. This is because U.S. interests in Latin America go back centuries, with the U.S. intervening in many of the region's countries over that time. As such, an evaluation of the possible ramifications of any actions would need to be done, as a number of those countries have felt a high degree of anti-U.S. sentiment. The benefit of such a psychohistorical analysis could provide the insight needed to avoid the political criticisms experienced after the 1961 Bay of Pigs incident.

THEMATIC ANALYSIS

Thematic analysis is another form of content analysis where a set of themes is distilled from the text. Ideas that present themselves in the reading are identified and given appropriate labels. The passages of text relating to a particular theme are marked and affiliated to the corresponding theme.

Using a manual system, this could mean photocopying the page text; cutting out the word, passage, or paragraph (i.e., using a pair of scissors); and placing it in a large envelope along with other passages that are identified. Each piece of paper (containing a word, passage, or paragraph) is referenced back to the text as one would do using, say, the Harvard referencing style, so that when the material is compiled into the final report, the analyst knows where it has come from (similar to note taking when writing an essay).

This type of analysis can be used in grounded theory research (as discussed in chapter 5) because it ideally lends itself to taking unstructured data and giving it organization. With grounded theory research, these themes then can be interpreted to explain phenomena (i.e., develop a theory about the issue under investigation).

THEMATIC METRICS

If a computer-based software package is used in connection to thematic analysis, then the same procedure takes place as discussed in terms of the manual system, but the software avoids all of the manual cutting, referencing, and storage issues, as this is done electronically. The advantages of using a computer-based solution are great, as the software will have other features, including word count and metrics such as the Flesch Reading Ease and the Flesch-Kincaid Grade Level tests (see the section later in this chapter). These types of analysis allow the analyst to use a blended approach, incorporating both qualitative and quantitative analysis.

Any number of documents can be analyzed using this technique, and in cases where, say, speeches are delivered by a foreign political figure in the country's local press, these public addresses can be analyzed over time in a quasi-longitudinal study. Alternatively, a number of author-known documents can be analyzed against a document of unknown or uncertain authorship with this method.

As an illustration, the Flesch Reading Ease and/or the Flesch-Kincaid Grade Level tests could be used where an analyst wants to gauge whether a speech was aimed at the local population of a developing country or whether its appearance in the daily newspaper was just a vehicle for projecting a message to international leaders. The use of these two tests could quickly yield results that suggest whether the speech under investigation had a reading ease score and grade level commensurate with the country's population or that it was much higher, perhaps suggesting it was aimed at a better-educated international audience.

Indexing is another way of identifying meaning in the text. Rather than identifying themes, indexing identifies keywords in context. This is best done with computer software. An exceptions dictionary is first set up in the software package where words such as *a, an, and, the, is, it, of,* and so on are flagged as exceptions, so when the software searches the text, it does not index these inconsequential words. All other words are then indexed.

These words appear in their context within the document. This allows the analyst to not only count the appearance of certain words—say, a repeated word like "infidel"—but also then tag the sentence or paragraph where these keywords appear so that they can be included in further analysis of a wider theme. Indexing is a technique closely tied to quantitative descriptive analysis. Descriptive analysis seeks to describe features of the text quantitatively as is done with numeric data—by describing the most frequently used words or phrases. This type of analysis ideally lends itself to a blended approach of both qualitative and quantitative techniques.

> "Not everything that counts can be counted, and not everything that can be counted counts."
> —Albert Einstein

Although content analysis has the advantage of having computer-based methods that allow the analysis of very large documents (and multiple documents), the technique is not without limitations. Firstly, the data need to be textual. If there is no source of textual data, no analysis can take place.

Caution must be exercised in terms of sampling bias as with other approaches. In the hypothetical case of the foreign leader's speeches in the local press discussed above, analysis would leave out all of the leader's speeches delivered by way of radio or television broadcast or in-person delivery to crowds assembled. The other limitation is that although software packages are useful in automating indexing, counting, and tagging text, these packages cannot interpret what words or phrases mean—it takes an analyst to do this word by word and phrase by phrase.

Flesch Reading Ease Analysis

The product of the Flesch Reading Ease analysis is a rating represented by a number on a scale between 0 and 100. The higher the score, the easier the text is to read. For instance, a document with the following Flesch Reading Ease scores would be interpreted as such:

- 90 to 100—very easy;
- 80 to 89—easy;
- 70 to 79—fairly easy;
- 60 to 69—considered to be what is generally termed plain English;
- 50 to 59—fairly difficult;
- 30 to 49—difficult; and
- 0 to 29—confusing.

The formula for calculating the score is: *readability ease* = 206.835 − (1.015 − *average sentence length*) − (84.6 − *average syllables per word*). Although most word processing packages will have this feature as part of its spelling and/grammar checker, it will be of interest to the intelligence analyst to understand how it is calculated. The average sentence length is determined by taking the total number of words in the document and dividing it by the number of sentences. The average number of syllables per word is calculated by taking the total number of syllables divided by the total number of words. Step by step:

1. Count all the words;
2. Count all the syllables;
3. Count all the sentences;
4. Calculate the average number of syllables per word;
5. Calculate the average number of words per sentence; and
6. Match the readability score.

Flesch-Kincaid Grade Level

The Flesch-Kincaid Grade Level analysis converts the Flesch Reading Ease score to a level equivalent to grade school rank (U.S.-based). The formula for calculating the score is *Flesch-Kincaid Grade Level* = (0.39 − *average sentence length*) + (11.8 − *average syllables per word*) − 15.59. For example, a score of 12 indicates that a person who has had a twelfth-grade education could understand the text contained in the document. Step by step:

1. Count all the words;
2. Count all the syllables;
3. Count all the sentences;
4. Calculate the average number of syllables per word;
5. Calculate the average number of words per sentence;
6. Multiply the average number of words per sentence by 0.39, and add this number to the average number of syllables per word, which is multiplied by 11.8;

7. Subtract 15.59 from the resulting number from step 6; and
8. Match the score with a U.S.-based school grade.

If a word processor or other software package is used to perform these calculations, note any limitations specified in the software, as some results may only report a grade level of 12 even though the grade level exceeds this figure.

SECONDARY ANALYSIS OF DATA

Secondary analysis of data is a quantitative approach that is closely aligned to content analysis. Secondary analysis of data relies on information that has already been collected. But rather than analyzing textual data, secondary analysis analyzes quantitative data for a second time; that is, an examination that is unconnected to the primary collection project.

As an example, intelligence analysts used secondary data analysis in the lead-up to the 1973 elections in France. At the time, decision makers in the United States were interested in knowing if a socialist-communist-left radical coalition was likely to form a government in France. Therefore, U.S. decision makers requested an intelligence assessment on the most likely outcome of the election. Intelligence analysts at the Central Intelligence Agency (CIA) were tasked with providing the assessment. Relying on existing data sets, CIA analysts used multiple regression analysis to gauge the impact various historical economic conditions had on the voting patterns of the left.[2]

Based on the results of this secondary analysis of existing data, analysts concluded that economic conditions did, in fact, impact elections but only in the absence of other important political considerations.[3] The use of existing data sets allowed analysts to predict that the domestic political factors affecting the elections would not be as powerful as those in previous elections—therefore raising the potential of a left victory. However, these political factors were also assessed as being strong enough to ensure that the required number of votes would *not* go to a left coalition. Arguably, such analysis was only made possible by using this technique.

Another example is the U.S. Department of State, which has no agents who engage in primary data collection. The Department of State's Bureau of Intelligence and Research therefore relies on data collected by other agencies to undertake its investigations.[4] This has been likened to the research conducted by Nobel laureates Doctors James Watson and Francis Crick, who, in 1953, discovered the double-helix structure of DNA by interpreting secondary data.[5]

The amount of data collected by governments around the world for social and economic planning is extensive. There are census data, crime statistics, social data, educational data, economic data, and consumer data, just to mention a few broad categories. Data are also collected by private corporations, think tanks, and a variety of nongovernment organizations for their own planning purposes, which are also available to outside researchers. Because these data are stored electronically, they can be imported into the software package being used by the analyst—say, a spreadsheet or a statistical software package such as SPSS. As was seen in the 1973 French election example, CIA analysts used data from several databases to conduct their multiple regression analysis.

Secondary data analysis is an efficient means of conducting research. Preexisting data sets alleviate the problem of a potentially lengthy collection phase that, in some cases, may have taken months or years to collect. It also means the data may be available at no cost or a modest fee, thus making it inexpensive as well. If a pretext is used to obtain the data set, it will not alert the target (or target country or corporation) that they are the subject of an intelligence operation. For instance, most small intelligence research studies are unlikely to have a budget large enough to conduct a national sample (or international as in the French case cited) because of the cost and time required, but by using a census data set, even a small research budget can gain considerable leverage.

Nevertheless, secondary analysis does have some limitations. In the main, analysts may find it difficult to gain a full appreciation of the problems encountered during the original collection or the errors inherent in the resulting data sets so that these limitations can be taken into account when manipulating the data during secondary analysis. It may also be a difficult task to link two or more data sets that have little in common— either by the structure of the database, the level of measurement (i.e., nominal, categorical, interval, or ratio), or units of measurement.

KEY WORDS AND PHRASES

The key words and phrases associated with this chapter are listed below. Demonstrate your understanding of each by writing a short definition or explanation in one or two sentences.

- Content analysis;
- Flesch-Kincaid Grade Level analysis;
- Flesch Reading Ease analysis;
- Psychoanalysis;
- Psychohistorical analysis;

- Psycholinguistic analysis; and
- Secondary analysis.

STUDY QUESTIONS

1. Explain how quantitative analysis can be conducted with unstructured data, say, in the form of a leader of a country's public speeches.
2. List five sources of secondary data.

LEARNING ACTIVITY

Select an editorial containing a few thousand words from a major national or regional newspaper (to ensure sufficient scope for several themes to be present). Using thematic analysis, highlight the various themes/subthemes discussed in the editorial. Using a pair of scissors, cut out these passages (or alternatively, number the themes *in situ*, e.g., 1 = political; 2 = social; 3 = educational; or, 4 = emotion; 5 = threats; 6 = lack of logic; and so on. The themes and how they are defined is part of the activity). Now summarize the editorial on two levels—the first being the overt message of the editorial and the second is the result of your thematic analysis. In the case of the latter, you may ask yourself whether the themes suggest anything beyond the simple message contained in the words.

NOTES

1. William Colby with Peter Forbath, *Honourable Men: My Life in the CIA* (London: Hutchinson, 1978), 337.
2. Susan Koch and Fred Grupp, "Regression Analysis: Impact of Economic Conditions on Left Voting in France," in *Quantitative Approaches to Political Intelligence: The CIA Experience*, ed. Richards J. Heuer (Boulder, CO: Westview Press, 1978).
3. Koch and Grupp, "Regression Analysis," 57.
4. See, http://www.state.gov/s/inr/ (accessed July 16, 2012).
5. James D. Watson and Francis H. C. Crick, "Molecular Structure of Nucleic Acids; A Structure for Deoxyribose Nucleic Acid," *Nature* 171, no. 4,356 (April 1953): 737–38.

11

᭍

Qualitative Analytics

A number of analytic techniques that can be used for unstructured data will be discussed in this chapter:

1. SWOT analysis;
2. PEST analysis;
3. Force field analysis;
4. Analysis of competing hypotheses;
5. Synthesizing matrices;
6. Fishbone analysis;
7. Morphological analysis;
8. Perception assessment analysis;
9. The third eye—trend prediction;
10. Timeline and key dates analysis;
11. Network analysis;
12. Telephone record analysis;
13. Event and commodity flow analysis;
14. Genealogical analysis; and
15. Financial analysis.

INTRODUCTION

One issue that is common to statistical analyses is "customer acceptance." Due to the abstract nature of the mathematical formulae of these techniques—like multivariate analysis—decision makers can feel troubled when asked to trust assessments based on such methods. However, this chapter describes a number of analytic methods that can be used to make sense of unstructured data in a range of intelligence settings that are easy to understand by decision makers and easy to employ by the analyst.

The reason for discussing a variety of methods is that each method tends to be problem specific. That is, some methods work better with certain types of issues, and others work better in other circumstances. Some methods are used for strategic assessments, while others are used for tactical or operational problems.

But just because an analyst has examined data using one of these methods does not mean that the results can be taken as absolute—analytic techniques are ways that allow analysts to form judgments (i.e., defensible conclusions) in a way that is transparent to the reader of their reports. Their analyses are able to be repeated in order to demonstrate the validity and reliability of the methods used (i.e., in keeping with the tenets of scientific methods of inquiry). The point is that analytic methods are not replacements for sound critical thinking or the application of professional judgment, but they aid both.

There is little doubt that the analysis of qualitative data enjoys no consensus about what technique is used.

SWOT ANALYSIS

Originally devised for corporate planning in the business community, analysis of strengths, weaknesses, opportunities, and threats (SWOT) is one of the most popular analytic methods used by intelligence analysts.[1] This is for two reasons:

1. it can be used with a variety of unstructured data (qualitative data from either primary or secondary sources); and
2. the focus of the research is not variable dependent—it can be either the target or the agency conducting the operation against the target.

The technique was devised for long-range business planning, but it can be applied to a variety of issues in the intelligence realm that are either tactical or strategic. Or it can be used to analyze information in order to build a profile or help understand the current situation. A SWOT begins with the analyst's defining the *end-state*, as it is called in a strategic setting, or *objective* if it is tactical.

It should be noted that the term *threats* is not used here in the same sense as *threat* is used in chapter 18 when discussing a *threat analysis*. In a security sense, *threat* is a person's resolve to inflict harm on another. However, in a SWOT analysis, *threat* is used to reflect detrimental factors—*risks, harms, dangers,* or *hazards*—not *threat agents*. This is because SWOT was originally devised by the business community for industrial and commercial forecasting; therefore businesses use *threat* in the generic sense. In a security environment, *threat* has a different meaning. So analysts should bear this distinction in mind when working with SWOT.

Using any one of several idea generation or data collation methods (discussed in chapter 6), the analyst populates each of the four quadrants of the SWOT matrix with the data (although a matrix typically displays a SWOT, SWOT analyses can be laid out in any way that is suitable for the analyst). For strategy assessments, it is advantageous to have a broad view of the issue and, hence, employ a multidisciplinary team approach (i.e., an ideas workshop) to consider each of the four factors. A tactical assessment could be done by an analysis based on the data collected in the lead-up to an operation or during an operation.

Once this has been done, it is a matter of assessing the factors one at a time and then cross-checking them for agreement (i.e., ensuring there are no contrary or paradoxical positions stated in different quadrants). Assessing can be done by asking hypothetical questions such as:

- In what way can the strengths be used to an advantage?;
- How can the weaknesses be shored up?;
- What is the best way to take advantage of each opportunity?; and
- What needs to be done to mitigate the threats (i.e., risks, dangers, or hazards)?

In a tactical setting, analysts can use the results of SWOT to examine a target's operating structure, method of operating, capabilities, financial base, and so on. A SWOT was used in developing target profiles in chapter 12 (with an example of a target profile involving a notional international criminal enterprise), and chapter 13 ("Tactical Assessments").

Ways of using the information contained within a SWOT can be generated based on combinations of the factors as follows:

Strengths/Opportunities. Ways that will use strengths so that opportunities can be realized.

Weaknesses/Opportunities. Ways to address weaknesses in order to provide relief so that opportunities can be followed.

Strengths/Threats. Ways that use strengths "offensively" to moderate threats.

Weaknesses/Threats. Defensive ways that will protect against threats.

See table 11.1 for an example of a SWOT analytical matrix.

Table 11.1. SWOT Analytical Matrix

	Analysis of Strengths, Weaknesses, Opportunities, and Threats	
	Supportive	*Detrimental*
Internal	*Strengths* are the attributes associated with the [issue/problem/agency/etc. under investigation] that are conducive to achieving the end-state.	*Weaknesses* are the attributes associated with the [issue/problem/agency/etc. under investigation] that are detrimental or may prevent achieving the end-state.
External	*Opportunities* are the conditions [legal/criminogenic/social/economic/political/psychological/etc.] that would assist in achieving the end-state.	*Threats* are the conditions [legal/criminogenic/social/economic/political/ psychological/etc.] that might be detrimental to the way the agency carries out its operations.

PEST ANALYSIS

If an analytic method could have a "cousin," PEST could be said to be related to SWOT. PEST is an acronym for political, economic, social, and technological factors (social factors could be couched in slightly broader terms, such as sociocultural, if desired; and technological factors can also include policy-related issues). These factors are usually the independent variables that are acting on the dependent variable. This technique has been used by the business community to assess the impact that these external factors might have on the organization or the market in which it operates. But, like SWOT, PEST can be used to an advantage by the intelligence community to assess a variety of issues under investigation.

Because PEST examines external factors, it is essentially half of a SWOT. But where the two differ is in the focus of the inquiry—PEST examines the environment in which the issue is positioned, whereas SWOT examines a dilemma or the actions of, say, a target. Viewed another way, PEST could be seen as the macro scene, and SWOT is the micro perspective. PEST is, therefore, used as the main method for analysts when conducting what are termed *environmental scans*. In this sense, PEST is more likely to be used for strategic analysis where the issues are complex.

PEST analysis can be conducted before a SWOT analysis (e.g., via an ideas workshop) to help identify issues, though it is less likely that a SWOT would be conducted before a PEST, as there may be issues raised in a PEST that would subsequently feed into a SWOT.

There are many variations of PEST. Some analysts add additional factors, thus modifying the acronym to variants such as STEP (PEST arranged differently—stay with PEST as it is universally known);

PESTELI (adding environmental, legal, and industrial factors); PES-TELOM (same as PESTELI, but instead of the industrial factor, it is substituted by organizational and media); STEEP (social, technological, economic, ethical, and political); and STEEPLED (social, technological, economic, ethical, political, legal, environmental, and demographic).

There is a view that these variants are not necessary and make the analysis overly complicated. Some scholars have argued that the four factors under PEST cover all of the issues that would arise out of an examination of the other subfactors. That is, the subfactors are contained in the main PEST factors. Nevertheless, any of these factors can be mixed and matched to suit the research project.

Finally, like the SWOT analysis, PEST has a simplicity that lends itself to ease of understanding while powerful enough to convey the results to decision makers. There are many examples, but two useful illustrations are shown in tables 11.2 and 11.3. One is a simple table and the other contains more detail. Both can be modified to suit most intelligence research projects or used according to the analyst's personal preference by adding, subtracting, or blending details found in these examples.

FORCE FIELD ANALYSIS

Force field analysis is a practical technique for examining the pressures that can be applied for or against a particular policy position, an operational tactic, or any other issue under investigation. It is a method that

Table 11.2. Simple PEST Analysis Template

Political	Economic	Social	Technological
List issues here	List issues here	List issues here	List issues here
A	A	A	A
B	B	B	B
C	C	C	C
D	D	D	D
E	E	E	E
F	F	F	F
G	G	G	G
H	H	H	H
Etc.	Etc.	Etc.	Etc.

Conclusions

Conclusions can be presented here based on the factors above.
These conclusions can be in point form or in narrative.

Table 11.3 Detailed PEST Analysis Template

	Comments and Observations	Impact Estimate High Medium Low Unknown	Timing 0-6mths 7-12mths 13-24mths 24+mths	Direction + Positive – Negative 0 Neutral	Rise/Fall > Increase < Decrease = Stable 0 Unknown	Import Critical Important Somewhat Not Very Not at All Unknown
Political	Include					
Factor 1	comment and	High	0-6mths	+	>	Not very
Factor 2	observations	High	0-6mths	+	>	Not very
Factor 3	here for each factor	Low	7-12mths	+	=	Somewhat
Economic	Include					
Factor 1	comment and	Medium	7-12mths	+	>	Not at all
Factor 2	observations	Low	24+mths	0	>	Important
Factor 3	here for each factor	High	7-12mths	–	<	Critical
Social	Include					
Factor 1	comment and	Low	7-12mths	–	<	Somewhat
Factor 2	observations	Low	13-24mths	–	>	Important
Factor 3	here for each factor	Medium	0-6mths	+	=	Not very
Technological	Include					
Factor 1	comment and	High	24mths	0	=	Critical
Factor 2	observations	Low	0-6mths	+	>	Important
Factor 3	here for each factor	High	7-12mths	+	>	Important

allows the analyst to form a judgment based on careful weighing of the pros and cons involved.[2]

For instance, a force field analysis can be carried out to weigh the possible success of a planned operation, or it can be used to gather *driving forces* in order to overcome or reduce the impact of *restraining forces*. In effect, the force field technique assumes a quasi-stable equilibrium point between these two sets of interests—driving forces and restraining forces—so that this "balancing" can take place. Thought of in another way, a force field analysis is analogous to an assets-versus-liabilities balance sheet. Force field analysis therefore lends itself to other variations, such as pros-cons-fixes and plus-minus-interesting—two methods that will be discussed shortly.

To illustrate how this works, consider a case of briefing decision makers about the likely impact of a newly proposed law that is intended to curb motorcycle gang violence. Based on the data obtained via the information collection plan, the analyst constructed a force field table that lists the pros and cons. A diagram that shows the conceptual layout of a table is shown in figure 11.1 and a generic table that follows this idea is displayed in table 11.4.

The analyst sourced her information from a brainstorming workgroup. This approach was particularly helpful, as the time frame of the issue being studied had a horizon of greater than six months. If the time frame was measured in years, then complexity increases and convening a multidisciplinary (and, perhaps, multiagency) group to brainstorm the issues would have been considered instead.

Using this example, then, using a data projector, the analyst displays and addresses the issues raised in the force field diagram, discussing the pros and cons, and finally presents the agency's preferred position. The preferred option, whatever it is, must directly address the original research question (i.e., will the proposed new law have an impact on curbing motorcycle gang violence?).

"There is no 'perfect' decision. One always has to balance conflicting objectives, conflicting opinions, and conflicting priorities. The best decision is only an approximation—and a risk."[1]

1. Peter F. Drucker, *Management: Tasks, Responsibilities, Practices* (Woburn, MA: Butterworth-Heinemann, 1974), 387.

When crafting recommendations, the analyst could also suggest changes to individual factors that, if implemented, could address the issue in favor of the driving forces, for it is the sum effect that the analyst is considering. To illustrate,

- Suppose there was a restraining force like this: new anti-gang legislation could inadvertently contribute to the workload of the court system, which in turn would result in trial delays.
- A recommendation that adds force to the drivers could be crafted as such: to prevent overlisting and trial delays when the new anti-gang legislation is enacted, prosecutors would need to work with the criminal courts' listing coordinators to manage trial lists.

Recommendations formulated along the lines of the second item could tip the balance from a position that suggests restraining forces dominate to one that sees driving forces considered. Viewed another way, force field analysis is a way to identify not only the forces for or against the issue under investigation but also to create a structure that allows the analyst to visualize ways of countering restraining forces. Step by step:

1. In a few words, describe the issue, plan, proposal, or policy option at the top of the table (see table 11.4). A separate table can be drawn up for each factor, attribute, policy, option, etc., that has been identified. Table 11.4 demonstrates just one such set of factors to illustrate this point;

Table 11.4.   A generic force field table showing pros and cons. Variations of this table can be made depending on the project and the issue under inquiry.

Why take a stand against international pirates.			
Driving Forces	*Score*	*Restraining Forces*	*Score*
To ensure the continuance of the international maritime trade route	+4	Large commitment of warships to interdict and support ships and personnel	−4
Maintain international commerce	+4	Area under pirate control is small compared to navigable sea lanes	−2
Project intolerance for wonton disregard for international law and conventions	+5	Pirates may not be sensitive to such posturing	−3
Ensure safety of crew on-board vessels	+5	Military personnel will be placed in harm's way during interdiction operations	−3
Option total: 18 − 12 = 6, suggesting moderate support for this policy position.			

2. In the left-hand column, list those attributes, options, or factors that can be considered driving forces in relation to the issue under investigation;

3. In the right-hand column, list those factors that are restraining forces;

4. Assign a numeric value to each force factor listed in the two columns. For instance, use an ordinal scale that ranges from, say, weak (+1) to strong (+5) for driving forces and for restraining forces, weak (–1) to strong (–5). In assigning a numeric value to each qualitative factor, analysts should be conscious that they do not bias the results by assigning values that reflect their own personal views, or the official view of the government of the day, or perhaps of one of the key decision makers (e.g., in order to curry favor). Projecting such bias into the method is unethical because it will artificially manipulate the analysis and use the scientific method simply as a guise for objective research. The safest way to assign the values is to achieve consensus through discussion with, for instance, a number of subject specialists that might take the form of a "judgment sample," "convenient sample," or other availability-based sampling technique.[3,4] If time and resources permit, the nominal group technique can be used to great advantage. Doing so removes any question of bias from the analysts (and the analytic unit that employs them).

5. Tally each column and add the two columns. If the total is a negative number, then the options regarding what the restraining forces are suggesting need to be considered carefully. If the number is positive, then the driving force options need to be considered. There is also the possibility that a zero result could occur or a weak (i.e., +1 or –1) result, suggesting that the direction may, on balance, be the way to proceed. But the analyst needs to apply judgment based on experience in such situations. Remember, this is just a method to guide thinking and reason; it does not reflect an absolute for any given situation.

Pros-Cons-Fixes

A variation of the force field analysis is the decision-making method known as *pros-cons-fixes*.[5] This variation starts by listing possible options for dealing with a problem. A corresponding note is placed next to each option, stating its positive attributes (this is usually done in table form but could be written up in a narrative). Then, all of the consequent negatives are listed, followed by ways the negatives can be overcome—these are the fixes. If no fix is available, then "no fix" is assigned to that option.

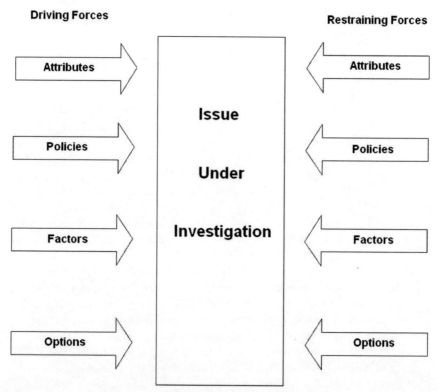

Figure 11.1. A conceptual representation of a force field analysis.

The final step is to relist all of the options but omit those with fixes. This is because options that can be fixed are seen as neutral; they are extraneous and can confound judgment. What are left in the table are a number of pros and cons that can be assessed more clearly.

Plus-Minus-Interesting

A quantitative version of the pros-cons-fixes technique is termed *plus-minus-interesting*. The steps in conducting this type of analysis are the same, except that instead of labeling the column *fixes*, it is labeled *interesting* issues; these can be either positive or negative. In addition, each issue comment (for all plus, minus, and interesting issues) in the table is rated on a Likert-type scale ranging from –5 (very negative) to +5 (very positive), with 0 as the neutral point midway, similar to the example cited for the force field analysis.

Once this is done, the analyst adds the ratings for all issues listed to arrive at a score for each option. A separate table is constructed for each option under consideration so that the total derived for each option can be assessed against the others being considered. Table 11.5 displays an example of one policy option relating to an issue involving international piracy.

ANALYSIS OF COMPETING HYPOTHESES

Analysis of competing hypotheses is a useful method to think about inductively construed theories in a rational way. It is an important method when the analyst is faced with several plausible propositions to explain the issue under investigation. Rather than being compelled to accept just one theory, analysis of competing hypotheses allows the analyst to evaluate all theories.[6]

Through this technique, the available evidence suggests the most plausible theory rather than having to decide on subjective factors (here, the term *evidence* is a generic term that also applies to arguments and the like). If there are insufficient data to draw a conclusion, the techniques can aid a new (or revised) information collection plan so that further or better data can be fed into the process. Managers of field operatives and other information-collecting assets can use the output of this type of analysis to task their resources more efficiently, saving valuable time.

Table 11.5. **Abbreviated Example of One Option of a P-M-I Analysis**

Option 1—Aggressive Stand against International Pirates		
Plus	Minus	Interesting
Boost domestic confidence (+3)	Some may see intervention as a risk to the lives of nationals living overseas (−2)	Media and press opportunities with other foreign leaders (+4)
Boost international reputation (+4)	Could be misinterpreted as a play for regional power (−2)	Joint operations (+2)
Secure safety of sea lanes (+5)	A further cost to the military (−1)	Valuable training for special forces (+2)
Ensure international trade (+3)	Military personnel to be overseas longer or more often (−1)	International law issues for intervention/interdiction (−3)
Option total: (15 + 5) − 6 = 14, suggesting strong support for Option 1		

Here are the steps involved in conducting an analysis of competing hypotheses:

- Draw up a matrix like the one shown in table 11.6, listing the various hypotheses across the top and important pieces of evidence down the left-hand column. Remember that it is not just the appearance of evidence that is always important; in some cases one might expect not to see a piece of evidence, or that key evidence has not shown itself. Each hypothesis and each piece of evidence does not have to be listed in full; a simple abbreviation of H1, H2 ... and E1, E2 ... will suffice. The analyst's descriptions in detail can be listed above or below the matrix as a reminder (as has been done in the heroin importation illustration below). At this point, the matrix is a quick way of bringing together all the information to form a clear picture.
- Working across the columns, assign a nominal value (+ or –) to indicate if each piece of evidence is consistent with the hypothesis or is inconsistent with the hypothesis.
- Tally the columns and consider whether the column with the most pluses should be advanced as the most likely hypothesis. The caveats placed on the conclusions drawn using force field analysis (see above) apply to this technique also.
- Just because one column may have the most "pluses" does not mean it is the best choice. Some pieces of evidence can and should carry more weight than others. In this regard, the process still requires the analyst to apply some degree of judgment before proceeding to advance this as the course of action.
- If sensitivity is an issue, analysts could consider using another scale, say, the ordinal scale, to attribute weight to each piece of evidence. For example, the use of double pluses (+ +) and double minuses (− −) could be added to the single signs discussed in this section, or a scale like that used in force field analysis could be adapted.

Although the use of the matrix is very useful, it is unlikely that any analyst will include it in the final report or briefing unless the audience is technically oriented (e.g., perhaps presenting the initial results to a peer group as part of a quality control process or validating the methodology).

Heroin Importation Illustration

To illustrate how this method can be used in practice, take for example the study of heroin smuggling into Australia in the early 2000s.[7] At the time it was observed by illicit drug users, police, and drug treatment professionals that heroin at street level was in short supply. There were four hypotheses put forward, and each was plausible. But which was the actual cause of the shortage?

The study explored these four possibilities, and a summary of the analysis appears below.[8] The study's research question is listed along with the various hypotheses and evidence. In table 11.6 the columns represent the corresponding hypotheses, H1 through H4. Evidence for these various propositions is listed in table rows. At the intersection of each column and row a sign appears that indicates whether that piece of evidence is consistent with the hypothesis or not.

Although in the study there was narrative discussion regarding each of these points, the table shown here is a convenient way of displaying at a glance the preponderance of evidence for the hypothesis that the shortage was caused by a Taliban-enforced reduction of Afghanistan-grown opium (i.e., H4).

Research Question
What are the likely factors that caused the heroin shortage in Sydney in 2001?

Hypotheses
H1—Recent seizures (at that time) by law enforcement agencies.
H2—The arrest of significant personalities in the supply and distribution chain.
H3—A severe water drought in the poppy-growing regions of Myanmar (Burma).
H4—A Taliban-enforced reduction of Afghanistan-grown opium.

Evidence
E1—Quantitative data about Australian law enforcement seizures.
E2—Elimination of unnamed and unspecified personnel.
E3—Data on crop production and rainfall in Myanmar.
E4—Quantitative data on drug production in Afghanistan.
E5—A 3,000-metric-ton reduction in Afghanistan-grown opium.
E6—Police intelligence of trafficking routes to Europe confirming a diversion of Golden Triangle heroin (destined for Australia) diverted to Europe to fill the Afghan void.

SYNTHESIZING MATRICES

In mathematics a matrix is an array of rows and columns that contain elements. Each element is located at the intersection of each row and column. For instance, a matrix comprising three rows and three columns can be populated by up to nine elements. These elements can consist of numbers, symbols, or other mathematical expressions.

Table 11.6. The Competing Hypotheses Matrix Regarding Prunckun's Study of Heroin Importation

	H1	*H2*	*H3*	*H4*
E1	+	−	−	−
E2	−	+	−	−
E3	−	−	+	−
E4	−	−	−	+
E5	−	−	−	+
E6	−	−	−	+

In intelligence research an analyst can use the mathematician's matrix to synthesize qualitative data. This is done in the same way as a matrix was used to aid analyzing competing hypotheses (see above). It is a quick way to understand relationships between the elements and at the same time provides a visual check of what information is available, or as a means of corroborating information, or any number of other research functions. It is also an easy way to present a briefing to decision makers.

By way of example, table 11.7 shows a matrix that synthesizes various sources of information regarding the likely capabilities of a notional terrorist cell to produce explosives. The first row across the top of the table shows a few sample questions an analyst could ask regarding the research question—"Does the group currently have the capability to manufacture explosives?" Although there are four listed here, any number of questions could be asked. At the left of each row is listed the confidential sources. These sources could be covert human as well as technical (e.g., CCTV and listening devices), as well as open sources. At the intersection of each row and column appear the corresponding data elements—in this case, there are qualitative descriptions.

Analysts can make variations to this simple analytic method by adding Likert-scale ratings to the qualitative data or by summarizing the strength of the data/observations in a column at the right of the table.

Although this example analyzes the capability of notional terrorist cells, the method can be used for a wide range of research questions in the same way matrices were used to test competing hypotheses (above). Synthesizing matrices can also be used to simply collate information. For a discussion of this, see the section entitled "Many-to-Many Matrix" in chapter 6.

FISHBONE ANALYSIS

Fishbone analysis is a qualitative method used to show cause and effect. It does this by identifying and then exploring the issues surrounding a

Table 11.7. Example of a Synthesizing Matrix for a Notional Terrorist Cell's Current Capability to Develop Explosives

	Group possesses existing knowledge	Group is attempting to acquire new knowledge	Group has necessary resources	Group funds to purchase resources
Source Alpha	Advised that two members are studying for degrees in chemistry	Unable to determine	Advised that safe house does not currently have materials to create explosives	Unable to determine
Source Beta	Unable to determine	Unable to determine	Unable to determine	Reports that all members of the group are well dressed, own vehicles, and live in luxury apartments. Has observed all members use tablet computers to access the Internet
Source Charlie	Confirms Source Alpha's information and advises that group's safe house has dozens of textbooks on chemistry	Reported that the group is trying to contact someone online who knows about how to stabilize the process of combining nitric acid and glycerin	Confirms Source Alpha's advice that group lacks resources for explosives at present	Advises that group appears well financed
Source Delta	Advised that chemistry textbooks were clearly present at safe house as well as laboratory glassware such as beakers, flasks, and test tubes	Noted that various members watch videos on the Internet regarding the use of ice to cool chemical processes	Source estimates that group has all laboratory equipment necessary to create a liquid explosive, but not the chemical ingredients	Observed one member logging on to her bank account via the Internet and was able to see the computer screen obliquely. Noted the bank balance was in five figures, but could not determine exact amount

problem that is under investigation. It can also be adapted by analysts as a technique to help manage the information collection process. In the case of the latter, see figure 3.1 in chapter 3 as an example of how a fishbone can be used to coordinate an information collection plan.

Applying the fishbone in practice follows this pattern: the analyst lists the problem to be investigated at the right-hand side of the diagram (the fish's head). He or she constructs the major bones of the fish by listing the major categories of information concerning the issue. Traditionally, the categories are machinery/equipment/devices, people, methods, and materials. Alternatively, the analyst can use categories such as policies, procedures, plant/equipment, and people. The categories used are not critical, as they merely act as a framework for analysis. They will be dictated by the issue being investigated.

Military intelligence analysts could use the nine order of battle factors to spotlight issues surrounding friendly, neutral, or enemy forces. These factors include (1) composition of organizational units; (2) disposition (location and deployment) of units; (3) strength of units regarding personnel, weapons, and equipment; (4) tactics that would be used by the units; (5) training at individual and unit levels; (6) logistics for unit supply; (7) combat effectiveness of the unit; (8) electronic and communications technical capability; and (9) miscellaneous background and supporting information about the unit.

From each of the major bones, minor bones can sprout to form a list of contributing issues. This is analogous to watercourses—small brooks and creeks that flow into larger streams, which, in turn, flow into rivers. Each issue contributes to form a larger problem. Fishbone analysis is aimed at identifying the problem's causes so that the analyst can suggest treatments. This analytic technique will not identify symptoms, because the symptom is the focus of the analysis (i.e., the issue under investigation).

When populating the diagram, the analyst may find that some topics contribute to more than one problem. This is good to note because when seeking a solution, it might be advantageous to address those issues that contribute to more than one problem. Analysts can use methods such as brainstorming and the nominal group technique to generate ideas for treatment.

To explain how this might be used in practice, figure 11.2 displays a template that could be populated with data relating to the piracy issue that was discussed previously. For instance, under the bone entitled

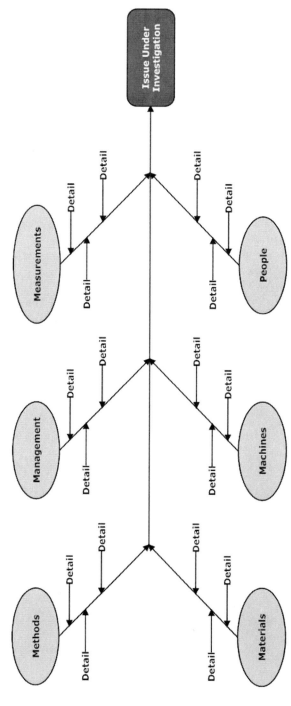

Figure 11.2. Typical template for conducting a fishbone analysis.

"methods" details such as single or small numbers of attack boats, and surprise attacks, could be listed. Under the bone entitled "materials" issues that might include fast speedboats, small arms, and explosives could be listed. And so on for each bone.

Each of the details listed along the major bones could have secondary bones with additional details. For instance, take the example of "surprise attacks" listed under the major bone of "methods"; additional details can be added, such as "dawn and dusk raids," as well as others, and this is applied to the other details cited in the diagram. The level of detail can be as scant or as detailed as is required to understand the issue under investigation.

By focusing on each of the "causes" that contribute to the "effect," analysts are able to put forward a set of intervening options for dealing with the problem. In the example being discussed here, analysts can formulate a number of possible interventions, each set addressing a particular aspect of the "cause." Suppose an analyst identified that attacks were isolated at dawn and dusk. This could be translated into a strategy of placing surveillance on alert around these times rather than on shifts over a twenty-four-hour period. Once all the possible solutions have been collated, the analyst could use the Pareto principle discussed in chapter 20 to prioritize those in the top 20 percent for implementation, should resources be limited.

MORPHOLOGICAL ANALYSIS

The term *morphological* is used in the intelligence context to explore complex policy issues. This is because one of the method's strengths is that it can be used to generate a large number of potential explanations for a phenomenon. Or it can be used to produce possible outcomes, treatment options, or causal theories for an event that has occurred, or may occur in the future.

In this regard, analysts can use the technique in place of brainstorming if time is a constraint or access to subject experts is not possible. Analysts can feed the output generated from this type of analysis into subsequent analyses, such as competing hypotheses. Step by step:

1. Deconstruct the issue under investigation to expose its component parts. As an example, take the research question: what means are available for controlling cyber weapons? Headings for analyzing this problem could be borrowed from the fishbone analysis—that is, either the categories of machinery/equipment, people, methods,

and materials or the categories of policies, procedures, plant/equipment, and people. Or analysts are free to devise their own categories.

2. Create a matrix to organize the headings associated with the research question. Table 11.8 is an illustration taken from Prunckun's study of cyber weapons control.[9] One will note that in the heading columns across the top of the table are listed the two control categories—prohibited and regulated. These are the categories that were identified in the deconstruction phase (step 1). In the rows descending down the page are the possible elements that comprise the category heading (this order can be reversed, if desired, with the headings down the left-hand column and the possible elements across each row).

3. Generate possible scenarios. There are several ways to generate scenarios. Starting with the first heading, the analyst selects an element from somewhere in the corresponding column and then moves across one and selects an element from the next row, randomly. This is repeated until one element is selected from each row. Another, more methodical way is to perform the selection in an orderly way that ensures every possible combination is included. In a two-by-six matrix this would result in thirty-six possible outcomes if they were combined in pairs, with another twelve possibilities if each of the elements that appear in the two columns are selected as stand-alone options. One can see that the matrix offers the analyst a mechanism for generating a large number of possibilities quickly, especially if other combinations were selected—for instance, one element from the first column and two or more elements from the second column, or vice versa.

4. Assess each of the explanations (i.e., output from step 3 above), and place them in descending order from most plausible to least plausible. Some of the possibilities generated may be quite illogical; nevertheless, the process affords analysts a range of possibilities that they can now consider as a part of a set of action items for decision makers. Alternatively, analysts can use the process to just identify options (selectively) that they may not have discovered using other methods and, as such, can explore through further research.

To demonstrate the use of his technique, the data in table 11.8 will be used to show the outcomes that are possible. Following the steps just outlined, an abbreviated list appears below. If the process was carried out in full, it would record all thirty-six permutations (i.e., 6 x 6, or 6^2), but for the purposes of this example, only a few are listed: 1A, 1B, 1C . . . 2A, 2B, 2C . . . 6A, 6B, 6C . . .

Table 11.8.   Morphological Analysis Matrix for Cyber Weapons Control

Prohibited	*Regulated*
1—Creation of cyber weapons	A—With existing criminal statutes
2—Possession of cyber weapons	B—With amendments to existing criminal statutes
3—Distribution of cyber weapons	C—Use existing licensing regime (e.g., for sporting guns)
4—Sale of cyber weapons	D—Create new licensing regime
5—Transfer of cyber weapons	E—For software weapons only
6—Use of cyber weapons	F—For hardware weapons only

One can see how many possible openings can be generated with just two categories. If other combinations were selected—e.g., one element from the first column and two or more elements from the second column, or vice versa—then many, many more possibilities could be generated: 1AB, 1ABC, 1DEF, 1ACE . . . 12AB . . . 23ABC, and so on.

Finally, in step 4 of the scenario generation guide above, an analyst could discriminate by means other than the two examples given—i.e., most plausible and least plausible. For instance, an analyst could discriminate by selecting combinations of factors that reflect low- or high-cost scenarios; or the scenario factors selected could discriminate by most/least likely to occur; or highest/lowest impact; or worst/best case scenarios; or scenarios that are able to be actioned/completed quickly; or even identify possible "black swan" events (i.e., highly improbable events with major consequences).

So if an analyst created a modest matrix of, say, six-by-six, the minimum number of scenario permutations would be 6^6 (i.e., $6 \times 6 \times 6 \times 6 \times 6 \times 6$) or 46,656. Obviously no analyst would have the luxury of time to map every possible permutation, but by using the suggestions above to discriminate by selective criteria, a "representative set" of scenarios could be developed.

Forced Relationships

A kindred method to morphological analysis is that of forced relationships. Although the technique follows a slightly different analytic approach, its intent is the same—that is, it will produce a large number of possible explanations or options that would not normally present themselves.

For instance, an analyst can use one of the idea generation techniques listed in chapter 6 to produce a list. Using the list, the analyst can then "force" a relationship between each of the items in the list with another item in the list. The end result will be a manifestly enhanced set of possibilities that may have been overlooked.

If we use the list of possible defensive counterintelligence strategies to protect office-based computer workstations (taken from chapter 21) as an example, the possible options for an effective system might look like this:

- Bolting the computer base unit to the workstation and locking the server room when technicians are not in attendance may be effective protection; or
- Positioning computer screens to prevent viewing from windows, doorways, or through glass partitions and allowing only trusted and qualified technical personnel to service or make modifications to a system may be effective protection; or
- Conducting electronic countermeasure sweeps at irregular intervals for bugs or wiretaps and shielding cables leaving the server room in metal conduit to prevent electromagnetic radiation, which could be intercepted, and to deter illegal tapping may be effective protection; and so on.

Other combinations, like those generated through morphological analysis, are possible too. Moreover, so is the ability to use the entire list. Nonetheless, the idea of forcing a relationship is demonstrated by way of this simple example. How the technique is applied is up to the analyst, the problem being considered, the time available, and the need to explore as many potential explanations or options.

PERCEPTION ASSESSMENT ANALYSIS

Perception assessment analysis allows the intelligence analyst to demonstrate relationships between actions taken by field operatives and the perceptions of those who will observe or experience those actions.

In psychology and the cognitive sciences, perception is defined as an awareness derived through sensory information. This technique enables operational managers to understand the possible impediments that implementing certain actions might give rise to because of perceptions. A matrix format displays the analytic results so that decision makers can understand how perception may become an impediment to policy implementation.

Taking an example from the military, an assessment might show how others in the operational environment—"enemy, civilian population, multinational, or coalition partners"—could perceive the dealings they have with friendly forces.[10] The matrix is ideal for this, as it lends itself to including other factors, such as criteria for determining "success."

But carrying out this type of analysis requires a more than average degree of understanding about the social and cultural issues that dominate the nation, the region, or the locality where troops are operating (as there can be both subtle and noticeable changes from one area to another). Nevertheless, if the analysts do not possess this knowledge, there is no reason why they cannot glean this information from subject experts (e.g., through in-depth interviews or focus groups). This knowledge is then used to assess the likely reactions the observers might have to actions by friendly forces. In a law enforcement context, friendly forces might be translated into a strategy to implement a neighborhood or business watch program, or other community-based crime prevention program.

Measuring perceptions is a difficult science at best. One could argue that there is a relationship between the magnitude of the physical stimuli (i.e., say, actions taken by friendly forces) and how a person perceives these actions. But this may not be the case in a sociocultural setting—what friendly forces view as a positive activity, the local population could perceive as insulting or disrespectful. These actions could cause a backlash against the actions being taken by friendly forces and the forces themselves.

Analysts can measure perception by several methods. Four methods are suggested by the U.S. Army,[11] and these are:

- Determine demographic and cultural factors that shape perceptions and reactions;
- Identify patterns and indicators from previous expectations and reactions in a society's history;
- Compare reported reactions to determine if they were based on real or perceived conditions; or
- Monitor editorial and opinion pieces of relevant newspapers for changes in tone or opinion shifts that can steer or may be reacting to the opinions of a society, organization, or group.

An example of a completed perception assessment analysis is shown in table 11.9. It appeared in the U.S. Army's interim field manual entitled *Open Source Intelligence*[12] and shows across the top the categories that need to be considered before action is taken. In each of the rows are presented the issues relating to the corresponding heading. The first three columns at the right of center are essential factual data, but the three rows to the left of center are interpretations of the data to the right and, as such, require expert subject knowledge in academic disciplines such as history, sociology, anthropology, theology, or political science.

"Population size and density, religion, social structure, and ethnic minority groups are important considerations in border security operations. Local customs may vary greatly. Troops must be educated to respect the local customs. Border security operations should minimize disruption of the customs, social activities, and the well-being of the population."[2]

2. U.S. Department of the Army, *FM 31-55: Border Security/Anti-Infiltration Operations* (Washington, DC: United States Government Printing Office, 1972), 7–2.

So take for instance this first condition—food. Here friendly forces are planning to distribute food to the local population who are struggling to feed themselves due to the insurgency. The intention is humane and honorable. However, the likely outcome is starvation due to the perception issues identified in this analysis.

This type of analysis provides transparency and replicability[13] and hence conforms to the scientific method of inquiry. It makes clear the facts and reasoning used to arrive at the final judgments. In doing so, the judgments are defensible.

Although this technique was discussed in a military setting, it can be used in other intelligence contexts—national security, law enforcement, business, and private. Police may use it in neighborhood gang intervention programs; businesses could use it for marketing goods and services in new markets overseas; and private sector intelligence could use it to gauge community reaction to concerns about privacy regarding the use of CCTV surveillance in certain situations.

THE THIRD EYE—TREND PREDICTION

A trend is the general direction that something is heading. This direction could be increasing, decreasing, maintaining a level path, or it could be any number of cyclical patterns or other trajectories. Predicting a trend can be useful in intelligence research in that it could provide insight for allocating operational field assets or other tactical tasks. Understanding trends is also helpful in long-term planning where preparations to deal with an anticipated situation may take months or years to put in place.

Trend prediction is a form of *futures research*. Futures research is a branch of the social sciences and is akin to history. Whereas historians

Table 11.9.   An Example of a Completed Perception Assessment Analysis

Condition	Cultural Norm	Friendly Force Action	Population Perception	Cause of Perception	Consequence If Unchanged
Food	Rice	Provided meat and potatoes	Inadequate and inconsistent	Practical (no experience with potatoes and cultural dietary rules on meat)	Starvation and riots
Armed civilian	All men carry weapons	Confiscated all weapons	Unfair and demeaning	Historical (previous experience with Western or military forms of government)	Risk of violence between U.S. forces and armed civilians
Government structure	Tribal	Establish military administration (hierarchical)	Tolerable as long as the authority fulfills needs		Loss of credibility and eventually control if needs are not met

study the past, futurologists study what might take place in the future. Arguably, it is part science and part art. It is science in that it uses a methodological approach that is transparent, but employs intuition, instinct, and perception.

There are many forms of futures research—enough to write a separate text on the topic. However, for an intelligence analyst who is tasked to conduct a study into the possible implications of a particular matter of concern, this technique is very handy and a basic understanding is all that is needed. It is called the *third eye* after the mystical reference to a person's invisible eye that provides perception beyond what can ordinarily be seen.

This third eye method does not require an enormous amount of time or elaborate preparations that are characterized by some futures methods. It only requires the analyst to ask questions of the situation that will provide a basis for a discussion of its implications for the future. Discussing the implications of the issue under consideration in the context of these questions can lead to an understanding of what might possibly develop and evolve in the future. For instance, by asking such questions as:

- Who would be most interested in this issue?
- What would these people or organizations do about it over the next three months, six months, or a year?

- Would the wider community be concerned about it?
- Would it only be of concern to a subculture?
- If they (the wider community or a subculture) were concerned, how concerned would they be?
- Would the wider community/subculture put pressure on political, religious, or community leaders to do something about it?
- What might those demands entail?
- How seriously would this leadership take such concerns?
- Could this issue be impacted by other developments happening concurrently?
- And so on . . .

This list is by no means comprehensive—it represents only a small number of questions that an analyst might ask when trying to predict a trend. Nevertheless, from just this handful of questions one can see that two or three pages of discussion could easily be generated. This discussion, based on existing factual data, could provide the mental stimulation needed to consider what the future could hold. Based on this, an analyst is in a better position to suggest options that could be actioned to deal with the issue.

TIMELINE AND KEY DATES ANALYSIS

A sometimes undervalued analytic technique for qualitative data is timeline analysis. It is akin to descriptive statistical analysis in that it describes in characteristic terms issues associated with key dates. The findings are displayed in a chronology table or a modified version.[14] The table provides analysts with a wealth of factual information that allows them to place proposed actions in sociocultural, political, and religious contexts to avoid clashes with, say, the local population, or on a wider scale, international actors.

Although what is discussed here is from a military perspective, timeline analyses can be used by law enforcers for issues like counterterrorism planning. These reports inform the decision maker how certain elements of a foreign population might react to friendly military force activity, or in a policing context, when and why local law enforcement might expect an attack by terrorists.

These timelines list such events as national, regional, and local holidays as well as religious and cultural events. In areas that have experienced political turmoil, public events and events that have symbolic political significance should also be noted.

It is common to include descriptions of the demographic makeup of the population and political developments relevant to the planned operations. Information that features in *basic intelligence* reference works like *The CIA World Factbook*[15] could also be considered for inclusion in this type of analysis—geography, economy, people, government, communications, and defense forces (see "Taxonomy of Intelligence Research" in chapter 2 for discussion on the meaning of *basic intelligence*).

An example of a timeline of key dates with pictograms is shown in figure 11.3.[16] Although this is an information-rich summary, presenting the table as-is to decision makers could prove overwhelming, and in doing so may risk the decision makers drawing conclusions that diverge from those at the center of the inquiry. For this reason, an analysis of key dates is likely to be used as a means of collating the characteristic attributes, and then used as a framework to analyze the data. The conclusions drawn from this process would be presented in a report in narrative form.

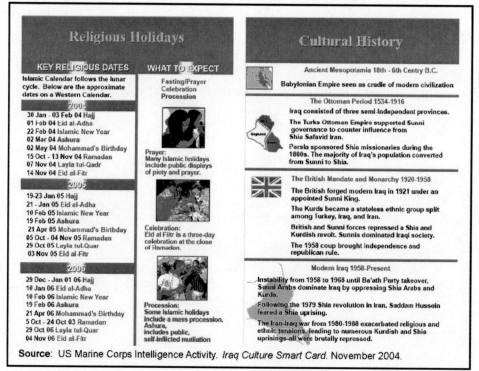

Figure 11.3.   An example of a timeline of key dates.
Courtesy of the US Marine Corps

A tactical assessment comes to mind as an example of the type of report where a timeline of key dates could be employed (see tactical assessments in chapter 13).

NETWORK ANALYSIS

If a research question focuses on the need to understand the relationships between two or more individuals, organizations, events, or other factors (or combinations), then network analysis can help make these associations clear. The relationships can be anything—social, business, financial, or even relationships that show abstract concepts such as influence, support, or mentoring.

The origin of network analysis is in the social sciences where scholars, like Moreno,[17] devised the use of two-dimensional diagrams to display relationships.[18] These were, and are, called *sociograms*, but in intelligence work analysts have called them *network analyses*.[19] Network analysis should not be confused with the closely related analytic technique known as *traffic analysis*.

Network analysis has a common bond with *traffic analysis*; the latter involves intercepting radio or telecommunications traffic between entities. Using analytic methods, patterns in these communications can reveal inferred meaning in the context of the issue under investigation. It is the exchange of the communication that is the subject of the analysis, not the content of the message. Times, days, frequency, duration, method of transmission, encryption method, and so on are the elements that are studied. As such, traffic analysis is an important methodology for situations where the targets are using encrypted radio transmissions.[3]

3. U.S. Department of the Army, *Fundamentals of Traffic Analysis (Radio-Telegraph)* (Laguna Hills, CA: Aegean Park Press, 1980). This is a reproduction of original Department of the Army *TM 32-250* (1948) and Department of the Air Force *AFM 100-80, Traffic Analysis* (1946), without changes. However, a glossary and an index were added by Aegean Park Press.

Network analysis is also called *association analysis* and *link analysis* because during the process, analysts use a matrix to show associations and lines to shown links.[20] Although these terms are used, this author is in favor of standardization with the term *network analysis*, as the other terms only describe part of the overall analytic technique. Because network analysis has a long history and tradition within the intelligence community,[21] analysts should stay with this term and avoid others.

Network analysis is a structured technique for unstructured data. It consists of plotting nodes that represent entities—circles indicate people, squares indicate companies or businesses, solid lines represent strong associations, dotted lines are used for weak or unconfirmed links, and the numbers along the lines count the number of contacts or interactions each entity has had with the other during the period of the study. Concepts such as *influence* can be represented by using arrows instead of lines.

The individual links that comprise the overall network can be described by the attributes of the associations between the entities. In a criminal context, these attributes might include routine pieces of information, like victims' addresses and phone numbers, their movements prior to the alleged offenses, and the offenders' modus operandi—all of which are analyzed in order to generate investigative leads, infer an organization's hierarchy (or lack of one), determine points of vulnerability/strength, and so on. Step by step:

1. Identify all entities. This can be done through other forms of analysis such as telephone record analysis or distilled from surveillance reports, notes or transcripts of interviews, telephone or e-mail intercepts, documents seized during raids, and so forth.
2. Assemble these data in an association matrix (see table 11.10 for an illustration). This is a standard matrix consisting of the same entities listed across the top and down the left-hand side. Where an entity intersects itself, the cell is blanked out. Where there is a known or confirmed association between two entities, a solid dot is entered into the corresponding cell for the two. If the association is suspected, a hollow dot is used, and a plus sign is used to denote that a person is a key individual in a company or organization.
3. Depict the entities as symbols on a diagram (the diagram can be a marker board, a flipchart, or a computer document), and draw the relationship lines (i.e., links) between them (see figure 11.4). Common symbols are circles for people, squares for corporate entities, and circles within squares for persons associated with an organization; solid lines represent strong associations, dotted lines for weak or unconfirmed links. Influential relationships use arrows pointing from dominant to subordinate. If using a computer program, it will have preassigned symbols for entities, for instance a telephone symbol for telephone numbers, a boat for watercraft, a car or truck for motor vehicles, silhouettes for people, and so on. There is no correct

Table 11.10. An Example of a Network Matrix

	Jack's Restaurant	Giacomo	Elizabeita	Rosa	Vladimir	Ignacy
Jack's Restaurant						
Giacomo	+					
Elizabeita	+	+				
Rosa	+	+	+			
Vladimir				+		
Ignacy	+	+	+			

way to display the entities on the chart. As intelligence projects differ, so too will each chart. The number and relationship between the entities will vary. As such, analysts will have to use their sense of artistic arrangement to create an exhibit that will clearly and easily show the viewer the relationships. If it is too "busy," it is likely to be confusing. If it is confusing, it defeats the purpose of presenting the data in this way.

TELEPHONE RECORD ANALYSIS

This is a variation to the matrix discussed in network analysis above. This technique is performed in the same way as network analysis, but instead

Figure 11.4. An example of a network diagram showing links between members of a fictitious European criminal enterprise.

of listing entities, the matrix lists the telephone numbers that were called on the left-hand side and the numbers that made the calls across the top. Where associations exist, a numeral is used to record the number of times that telephone called the other.

Taking these data, analysts construct a diagram exactly as they did for network analysis, but arrows are used to show the numbers making the calls to other numbers. The frequency of calls made is shown as a numeral alongside the arrow. Landline telephones, pagers, cell (mobile) telephones, fax machines, and computer modems can all be part of this analysis.[22]

The technique streamlines an otherwise labor-intensive process, especially if there is a large volume of data (e.g., obtained from telephone bills or pen registers). If this technique is used in conjunction with a network analysis of entities, the findings could help identify targets for installing covert listening devices or for conducting physical surveillance.

EVENT AND COMMODITY FLOW ANALYSIS

Event flow analysis is a technique that is used to clarify situations that comprise multiple events and take place over a period of time—hours, days, weeks, months. These situations are usually characterized by complex simultaneous or closely related historic events that would be confusing to understand unless ordered sequentially in time.

Analysts can also use this technique to give structure to qualitative data in order to understand the flow of commodities—for instance, illicit drugs, stolen goods, importation or exporting of restricted/prohibited items, sales of weapons or their components, explosives, precursor chemicals, technology, and the like.

Therefore, an event/commodity flow analysis is a diagrammatic depiction of the chronological events/movements. Once visualized in this way, analysts can draw inferences to generate further investigative leads or to test a hypothesis. Step by step:

1. Identify all events associated with the issue under investigation. Analysts can obtain these data from crime scene examinations or, as with network analysis, through the distillation of surveillance reports, notes or transcripts of interviews, telephone intercepts, documents seized during raids, and so on.
2. Assemble these data in a chronology table. This is a similar process to the timeline table produced in the analysis of key dates (above).
3. Depict each event as a symbol on a chart (the chart can be a marker board, flipchart, or computer document) and draw a progress line between them. Times that separate the events can be shown as vertical dividing lines at uniform time periods. If two or more events occur at the same time, then the progress lines joining the events will split and connect to the simultaneous events. These lines will converge again as they move to the next event(s).

GENEALOGICAL ANALYSIS

Genealogy is the study of families by searching their lineages back through history. From time to time, intelligence analysts may be called upon to provide a descriptive analysis of the extent of a target's family relationships. A case in point is the late dictator of Iraq, Saddam Hussein, where a detailed understanding of his family's relationships aided in his capture in December 2003.[23]

Data used in this type of analysis can come from many qualitative sources: oral histories (of relatives); personal historical records (e.g., family holy books); birth, death, and marriage records and newspaper notices of same; grave and burial records; census records; and military and other government records. Analysts can use any source of information that shows kinship or pedigrees. Contrast genealogy with family history research—the former is a study of kinship, while the latter is a more detailed investigation into the lives and history of the family members.

Analysts display genealogical research findings in diagram form, as the visual presentation of the associations is easy to follow. They also use written narratives, but unless the number of kinships is few, such narratives can confuse the reader (though analysts can use an indented outline to show generations more plainly).

Analysts can draw pedigree charts in a number of ways, including freehand and with preprinted templates. But the use of a computer package to collate and display the relationships is the most efficient method. These packages typically have other features for producing reports, including the ability to generate group sheets for individual families; descendant or ancestor charts; individual relationship charts and individual summaries; lists of anniversaries (e.g., birth, death, marriage); and if photographs of the family members are available, these can be incorporated into the genealogical database and displayed in the pedigree diagram. As an illustration, an invented family that is involved in a hypothetical international organized criminal enterprise is shown in figure 11.5.

FINANCIAL ANALYSIS

Although not strictly a qualitative method, financial analysis is considered here, as it takes unstructured data and gives it structure so that conclusions can be drawn. Granted, these data items are numbers and not words, concepts, themes, or other pieces of information without form; a person or organization's financial position can be of the ilk—physical assets, income from a number of sources, cash, investments of various descriptions, and the like.

The single most useful analytic technique for the non-accountant analyst is net worth analysis. In most agencies, financial analysis is conducted by a qualified forensic accountant because of the degree of knowledge it requires to understand the processes and procedures to reconstruct the target's financial position.

Nonetheless, an intelligence analyst can conduct a preliminary analysis of the target's net worth using the method described here. If the analogy of a paramedic and a surgeon is used, the intelligence analyst would be the paramedic performing the first-stage analysis while calling in the surgeon (perhaps a forensic accountant) to perform the more intricate analysis once the central issues have been identified.

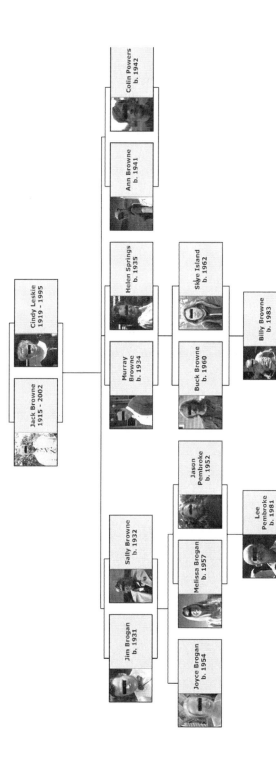

Figure 11.5. An example of a photographic genealogical analysis of a notional crime family.

Net Worth Method

Net worth is an indirect method of assessing the target's income and, therefore, is a handy technique in situations where the analyst may have come across information that suggests some form of illegal enterprise or secret deals (e.g., arms trafficking). It could also be used to assess a target's suitability to an approach by a field operative who is hoping to recruit the target as an agent (e.g., an offer of financial assistance).

Net worth is simply the difference between the target's assets and liabilities. If the analyst conducts a net worth analysis over a period of time, say, for the end of each financial year, he or she can compile a picture as to whether the target is growing in worth or is experiencing losses and what the magnitude of these gains or losses might be. The formulas for calculating net worth are as follows, step by step:

1. Assets – liabilities = net worth
2. Net worth – prior year's net worth = increase or decrease in net worth
3. Net worth increase (or decrease) + living expenses = income
4. Income – funds in known sources = funds from potentially illegal sources[24]

KEY WORDS AND PHRASES

The key words and phrases associated with this chapter are listed below. Demonstrate your understanding of each by writing a short definition or explanation in one or two sentences.

- Competing hypotheses;
- Driving forces;
- End-state;
- Environmental scan;
- Event and commodity flow analysis;
- Financial analysis;
- Fishbone analysis;
- Force field analysis;
- Genealogical analysis;
- Morphological analysis;
- Network analysis;
- Net worth method;
- Perception assessment matrix;
- PEST analysis;

- Restraining forces;
- Sociogram;
- SWOT analysis;
- Telephone record analysis; and
- Traffic analysis.

STUDY QUESTIONS

1. What are the four quadrants in a SWOT analysis? Describe each, and explain what type of data the analyst would seek to populate each.
2. Compare and contrast SWOT with PEST. Discuss when an analyst might use one technique over the other and why.
3. Describe the variations that can be applied to PEST.
4. Summarize the steps in creating a network analysis.
5. Discuss situations when an analyst might use event flow analysis and commodity flow analysis.

LEARNING ACTIVITY

Suppose you have been tasked to construct a pedigree chart for the dictatorial leader of Country Q. Your findings will form part of a psycholinguistic analysis of him. Using hand-drawn lines or a software package, create a pedigree chart. For the purpose of this learning activity, use your own family members as a way of indicating your skill in creating the chart. Seek as many different sources of data as possible to simulate a real-world project. If stumbling blocks are encountered in obtaining data about family members (as they would be with a real target), consider how you could collect this information from other sources. List the possible alternative sources and methods of acquiring these data (e.g., using an information collection plan format).

NOTES

1. The technique is sometimes referred to as SLOT analysis because the L stands for *limitations* rather than the W for *weaknesses*. In both cases, the same idea is represented.
2. This technique is based on the method attributed originally to Kurt Lewin's discussion which appeared in Dorwin Cartwright, editor, *Field Theory in Social Science: Selected Theoretical Papers* (New York: Harper & Row, 1951).
3. Gary T. Henry, *Practical Sampling* (Newbury Park, CA: Sage, 1990), 17–20.

4. Gennaro F. Vito, Julie Kunselman, and Richard Tewksbury, *Introduction to Criminal Justice Research Methods: An Applied Approach*, second edition (Springfield, IL: Charles C Thomas, 2008), 125–28.

5. Former analyst with the Central Intelligence Agency, Morgan D. Jones, *The Thinkers Toolkit* (New York: Three Rivers Press, 1998), 72–79.

6. Richards J. Heuer, *Psychology of Intelligence Analysis* (Washington, DC: Center for the Study of Intelligence, Central Intelligence Agency, 1999).

7. Hank Prunckun, "A Rush to Judgment?: The Origin of the 2001 Australian 'Heroin Drought' and Its Implications for the Future of Drug Law Enforcement," *Global Crime* 7, no. 2 (May 2006): 247–55.

8. Prunckun, "A Rush to Judgment?", 247–55.

9. Hank Prunckun, "'Bogies in the Wire': Is There a Need for Legislative Control of Cyber Weapons?" *Global Crime* 9, no. 3 (August 2008): 262–72.

10. U.S. Department of the Army, *FMI 2-22.9: Open Source Intelligence* (Fort Huachuca, AZ: Department of the Army, 2006), 4–16.

11. U.S. Department of the Army, *FMI 2-22.9: Open Source Intelligence*, 4–16.

12. U.S. Department of the Army, *FMI 2-22.9: Open Source Intelligence*, 4–16.

13. Some scholars prefer the term *reproducibility*. Both terms are perfectly acceptable, as it is the intent of science to be able to verify research results, detect intellectual fraud, and understand limitations of various research approaches, data, and so on. The view taken here in this book is that both terms reflect this intent.

14. Marilyn Peterson, *Applications in Criminal Intelligence: A Sourcebook* (Westport, CT: Greenwood Press, 1994), 36.

15. Central Intelligence Agency, *The CIA World Factbook, 2012* (New York: Skyhorse Publishing, 2011).

16. U.S. Department of the Army, *FMI 2-22.9: Open Source Intelligence*, 4–18.

17. Jacob L. Moreno, *Who Shall Survive? Foundations of Sociometry, Group Psychotherapy, and Sociodrama* (Washington, DC: Nervous and Mental Disease Publishing, 1934).

18. John Scott, *Social Network Analysis: A Handbook* (Newbury Park, CA: Sage, 1991).

19. Henry Prunckun, "The Intelligence Analyst as Social Scientist: A Comparison of Research Methods," *Police Studies* 19, no. 3 (1996): 67–80.

20. International Association of Law Enforcement Intelligence Analysts, *Successful Law Enforcement Using Analytic Methods* (Alexandria, VA: IALEIA, 1997).

21. Francis Ianni and Elizabeth Reuss-Ianni, "Network Analysis," in Paul Andrews and Marilyn Peterson, eds., *Criminal Intelligence Analysis* (Loomis, CA: Palmer Enterprises, 1990).

22. Marilyn Peterson, "The Context of Analysis," in Paul Andrews and Marilyn Peterson, eds., *Criminal Intelligence Analysis* (Loomis, CA: Palmer Enterprises, 1990).

23. U.S. Department of the Army, *US Army Field Manual 3-24/ Marine Corps Warfighting Publication 3-33.5* (Chicago: University of Chicago Press, 2007), 323.

24. Leigh Edwards Somers, *Economic Crimes: Investigating Principles and Techniques* (New York: Clark Boardman Company, 1984), 99.

12

⟡

Target Profiles

The topic discussed in this chapter focuses on a particular type of operational-orientated intelligence report, and will cover the report's:

1. Introduction;
2. Background;
3. Key parts of a target profile;
4. Terminology; and
5. An example of a target profile.

INTRODUCTION

A *target profile* is a type of an intelligence report that summarizes information about a specific target. It is an operationally focused report that is considered a short-form report.[1] Joby Warrick in his book on a CIA double agent operation describes the role a target profile plays: "Like an artist assembling a giant mosaic, [the targeter can] summon bits of information from wiretaps, cell phone intercepts, surveillance videos, informant reports, and even news accounts [as well as other open sources] . . . to develop a profile that the agency's spies, drone operators, and undercover case officers [can use operationally]."[2]

A target can be an individual or group, but it could also be a company or organization. In the case of the latter, the report may be entitled a *criminal business profile, terrorist organization profile,* or similar name. A variation to the target profile is the *problem profile*—this is a report that focuses on an issue, not a person or organization. An example could be a series of crimes occurring in a geographic "hot spot." In any case, the target can be either the subject of a current inquiry or an emerging target for a proposed investigation. The profile report summarizes what is known about the target and, in doing so, identifies information gaps, which feed into a collection plan for additional data.

BACKGROUND

Target profiles often provide field operatives with a range of options regarding possible intents (i.e., hypothesis derived from inductive reasoning) or possible ramifications if the target continues the activity at the center of the report's concerns (e.g., risk assessment). In doing so, a good report will prioritize the need to make further inquiries about the target in ranked order with other targets under investigation so that intelligence unit managers can allocate resources.

A target profile consists of several sections that are arranged to "tell a story": background, personal details, criminal record (in the case of a law enforcement target), physical environment, analysis, and target planning. There may also be attachments appended to the end of the report.

Even though the different components that comprise a target profile are discussed in this chapter, for clarity, an example of a completed target profile appears at the end. This example relates to a hypothetical criminal enterprise that, for the purposes of the exercise, was suspected of operating in several European and Middle Eastern countries as well as in India.

> *Context* is the background to a problem or situation. It allows analysts to make sense of why an event is happening by allowing them to explain key contributing factors.

KEY PARTS OF A TARGET PROFILE

Introduction

This section provides brief details about the legal basis or agency policy which gives the analyst authority to compile the report, and the name of the authorizing officer (e.g., the intelligence unit's manager), date, and file number for audit purposes.

Background

The background section of the target profile contains a short statement of the report's aims or objectives, scope, and a brief description of how the target fits into the broader picture—the context (e.g., motorcycle gangs in the region).

Personal Details

This section provides a descriptive analysis of the target (sometimes referred to as the *person of interest* or abbreviated as POI). It will contain a biography and physical description of the target and, if available, a photograph. Any aliases used by the target are included along with a description of the target's usual or last practiced occupation, business address, business affiliations, and so on. Other pertinent facts, such as driving licenses, trade licenses, and vehicles owned or regularly driven (in the case that they do not own a vehicle), appear in this section.

The personal details section can also contain details about the target's social and/or psychological functioning. This could take into account information about places the target frequents, personal friends, close family connections, and business associates.

Business Details

A range of information about the target's own business or the businesses they are associated with appears here. Rather than being prescriptive about what should be included in this section, suffice it to say that any detail that bears directly on portraying any of the elements of the *Kipling method*—what, why, when, how, where, and who—associated with the aim or goal of the report should be included. By way of example, in a money laundering case, details should be included on off-shore banks, electronic fund transfer arrangements, businesses and nonprofit organizations in the cash flow chain, key personnel, financial information, and company data from the government agency responsible for registering companies in the jurisdiction, and so on.

I keep six honest serving-men (They taught me all I knew); Their names are What and Why and When and How and Where and Who.[1]

1. Rudyard Kipling's poem, "I Keep Six Honest Serving-Men." See, for example, *Animal Stories* (Cornwall, UK: The House of Stratus, 2011), 134. The method is also known among journalists and law enforcement investigators as the *Five Ws and H* (and is sometimes abbreviated as 5W1H).

Criminal Record

In addition to outlining the target's criminal history, this section could contain a list of court appearances, how they responded to previous probation or parole orders, and, if currently on bail, bail conditions and reporting arrangements. There may be intelligence from a prison or jail intelligence unit that could prove insightful—if available, consider including it.

If the target has prior convictions or has been arrested for violence or firearms offenses, these details need to be highlighted as an occupational safety issue for fellow officers who might have to deal one-on-one with the target. Further, the types of crimes or the frequency of offense may be relevant as to why the target profile is being developed, and therefore, it too should be a feature. If agency records note the target's modus operandi, then this information will be another important aspect for inclusion.

With regard to corporate and private sector intelligence, where practitioners do not have legitimate access to these data, analysts may only be able to obtain this type of information from public databases. Newspapers, both hard copy and online, report on criminal activities that take place at local, state or provincial, and national levels. These reports contain the offenders' names, their offenses, court convictions, and probation and parole details. This is because this information forms part of the public record. These data are suitable for inclusion in this section as long as it is recorded what is proved as fact (e.g., in a court of law) and what is alleged by a law enforcement agency. Not all criminal matters are reported in newspapers, but public databases like this are important sources of information for corporate and private intelligence analysts.

Criminal Associates

Sections like this can be added to the target profile in order to provide additional facts about the target. Here details of the target's associates who have criminal records are listed. Information relating to their arrests and/or convictions is listed in reverse chronological order. If the target's associates have a probation record, a prison file, or a parole dossier, those details can be included in this section, or for clarity, they can be provided under a new section heading to correspond to that information.

Physical Environment

Observations made by surveillance operatives appear in this section of the profile. Details about the target's physical space complements what is known about his or her social and psychological makeup (e.g., arrests and convictions can be considered external manifestations of their psyche).

Moreover, surveillance operatives may have noted some of the ways the target practices countersurveillance or has put in place other intelligence countermeasures. Surveillance reports may have also commented on whether these precautions are conducted regularly, on an *ad hoc* basis, and whether they appear to be effective.

Analysis

This is the "engine room" of the report. The material that appears in the sections just discussed is descriptive in nature. However, in this section, the analyst applies one or more analytic methods so these data provide insight. It is sometimes described as the "so what" section. That is, what do all the previous facts mean, and what implications do they have? Depending on the issue or allegation, it might be as simple as positioning the target within the wider criminogenic landscape, or it might be as sophisticated as detailing where the target fits within a transnational crime syndicate, terrorist group (or offshoot), espionage ring, military establishment, and so on.

This is done via analysis; analysts cannot rely on "a hunch," "gut feel," or base it on "experience" or "belief." If a conclusion is drawn, then analysts must be able to demonstrate how they arrived at this finding. For instance, if an assessment centers on the risk the target poses to the community (or field operatives, etc.), then a risk analysis must be conducted.

How elaborate this analysis is depends on the facts and the issues surrounding the target as well as the aim of the report. But generally, for the purposes of a target profile, it can be a simple, straightforward table examining the likelihood and consequences of key hazards (i.e., the sources of risk).

If business and financial information is presented in the business section of the report, the analyst could then present some form of financial analysis or network analysis of the key personnel involved.

Even if the data are "thin," at a minimum an analyst can conduct a SWOT analysis that examines the strengths, weaknesses, opportunities, and threats posed by the target. Such an analysis can highlight missing data (i.e., intelligence gaps) and form the basis for recommendations for additional data collection via a new information collection plan.

The analyst can insert additional analyses depending upon what data were described in the preceding sections, but one important analysis that the analyst should always include is that of countermeasures (if any). A simple force field analysis of the precautions taken by the target when using landline and mobile telephony, voice over Internet (VOiP) communications, data transmission (including e-mail), two-way radio, and satellite

"It's no sin to admit an intelligence gap. . . . By admitting to the unknown, we may get someone's attention and initiate some seriously needed collection action. Your work might make a positive contribution by calling someone's attention to an intelligence gap of possible consequence."[2]

2. James S. Major, *Writing Classified and Unclassified Papers for National Security* (Lanham, MD: Scarecrow Press, 2009), 8.

telephony would be helpful for planning how field operatives are tasked in the future. If these are missing, it needs to be highlighted.

Include also observations made when the target makes face-to-face contact with "significant others"—for instance, does the target practice countersurveillance techniques in these situations? Are codes, ciphers, or encryption used when they communicate?

Target Planning

Having presented the known facts and analyzed these to gain an understanding into the target's activities, this section takes the insights developed and answers the question of "where to from here?" or "so what do we do now?" The planning section usually starts off with a short narrative summarizing the findings in the context of the overall aim and then discusses practical and realistic ways of addressing the problem.

This is usually done by providing a range of policy or operational options—from the "do nothing" option to an option that could be described as a "gold standard." Unless there are large political pressures being applied because of the actions of the target, the Rolls-Royce option is normally outside most agencies' budgets, so one of the middle options is going to be more attractive (acknowledging that there will be some trade-offs in effectiveness and/or efficiency which should be noted in the report). The bottom line is to present decision makers with a way of stopping, disrupting, or reducing the illegal, harmful, or otherwise detrimental effects of the target's activities.

The do-nothing option, although at first glance it appears an option that would be dismissed outright, could be a viable choice where further time is required to either gather critical pieces of information or where all necessary data are in hand and arrest, capture, seizure is the next and final step. If more information is required, then the analyst must be mindful of the costs that will be incurred in gathering these data and the likelihood that such intelligence gathering will benefit the analysis and improve the recommendations of the report in a material way.

Attachments

Because a target profile is usually a short, sharp, and focused report, attachments should be kept to a minimum. A few examples of what might appear in an appendix are a map, a photograph, or an organizational chart showing complex relationships that would be confusing if they appeared in a narrative form in the body of the text.

TERMINOLOGY

In practice, some terminology used in the intelligence arena is applied inappropriately. As practitioners, it is important that we use key terms correctly; otherwise, our reports will lack credibility. Following are a few words that typically cause difficulties. Next to each term appears a definition. No doubt there are other definitions, but what is important in considering these is that analysts focus their thoughts on writing precisely.

- **Allegation.** An accusation yet to be substantiated.
- **Analysis.** Systematic examination that follows some scientific, mathematical, or logical procedure or process.
- **Assumption.** Something that is considered to be true without proof.
- **Believe/Belief.** Synonymous with *faith*, and hence not a recognized intelligence methodology. Suggest that the word *consider* is used. *Consideration* is based on fact and reason.
- **Conclusion.** A judgment or finding.
- **Consideration.** Deliberation, a process of long, careful thought.
- **Evidence.** Something that provides an indication that something exists or contributes to the process of proving the truth of something.
- **Fact.** Something that can be observed or experienced through one of the five senses.
- **Hypothesis.** A theory, an explanation for something, which is then used as the basis for examination or investigation. Speculation. Conjecture, guesswork.
- **Inference.** A theory constructed from data that have been subject to analysis.
- **Intelligence.** "Insight" expressed in the form of a product (e.g., report, target profile, target assessment, estimate) and the process that produces such a product.
- **Investigation.** The assembling (i.e., pre- or during) or reconstruction (i.e., post-) of facts surrounding an event.

- **Postulate.** Something that is considered to be true and forms the basis of a theory.
- **Probability.** Likelihood, chance of something happening. Statistical confidence is the probability that a statement is true. If probability is not based on analysis, it is conjecture. (The use of words such as "will" conjure the idea of 100 percent certainty—avoid saying "something will" unless probability analysis confirms this.)
- **Proof.** The legal process of introducing a fact into evidence with the intent to establish the truth of something.
- **Standard of proof.** Beyond reasonable doubt (criminal) and on the balance of probabilities (civil).
- **Suspected.** Alleged.
- **Truth.** A subjective notion. For instance, it was once viewed that the earth was flat, and this was the accepted truth at the time. To have questioned this truth was heresy. In the criminal justice setting, truth is derived via a judicial process of proofing evidence in accordance with the rules of evidence, criminal procedures, and precedent.

Finally, different words mean different things to different people. Adjectives that invoke some level of sensationalism should be avoided in an intelligence report. The best way to do this is to "describe" or "explain," but do not excite. For instance, what an analyst might see as "extreme" may be "insignificant" to the reader.

Avoid the use of personal, subjective, or unproved ideas and ambiguous language in intelligence reports. Statements made need to be supported by evidence.

AN EXAMPLE OF A TARGET PROFILE

Introduction

Date:	February 28, 2014
Authorizing Officer:	Director, National Crime Analysis Agency (NCAA)
Crime Analysts:	International Organized Crime Group—Delta X-Ray
In the matter of:	Section 50, Anti-People Smuggling Act, 1992
File:	I-0043/2014

Background

INTERPOL has requested the National Crime Analysis Agency to assist it in its investigation of an allegation of people smuggling and it is suspected that Lucien *DEWIRE* may be implicated in this criminal enterprise. The alleged criminal enterprise is suspected to operate in several countries, including France, the United Kingdom (UK), the United Arab Emirates (UAE), and the Republic of India.

Personal Details

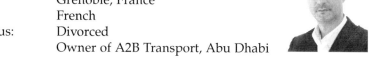

Name:	Lucien *DEWIRE* (AKA "The Wire")
DOB:	1 April 1983
Born:	Grenoble, France
Citizenship:	French
Marital Status:	Divorced
Occupation:	Owner of A2B Transport, Abu Dhabi

Business Details

- 2000 to 2002—telemarketer
- 2002 to 2005—insurance salesman
- 2005 to 2008—used car salesman
- 2008 to 2010—printing industry salesman
- 2010 to present—partner and co-owner of a trucking company in Abu Dhabi, United Arab Emirates, that trades as *A2B Transport*

Criminal Associates

(1) Alain *PRESSCOT* (AKA Albee *PRESS*), French national and a person friendly with the Mignon Syndicate (Marseilles, France). The Mignon Syndicate is reported by INTERPOL to be involved in people smuggling into the UK. *PRESSCOT* meets socially with members of the Syndicate when he visits Marseilles. University educated with a Bachelor of Business Degree. He currently owns a fleet of container trucks in Abu Dhabi. No criminal record.

(2) Gunther *LETZ* (AKA "Grunter"), German National, antecedents consisting of arrests for assault with a deadly weapon (switchblade knife) and possession of an illegal firearm (M1 assault rifle). Ex-German soldier with combat experience in Afghanistan (reconnaissance) and has trained "security guards" (described in the press as "mercenaries") throughout Africa for various mining companies. He is reported by INTERPOL to be a member of the Mignon Syndicate, but owns a charter boat company in Port Sudan and has several oceangoing vessels. INTERPOL reports that he is under investigation by it for people smuggling.

Table 12.1. Criminal Record

Year	Offense	Penalty
2010	Arrested for alleged money laundering $80,000	Charges dropped, insufficient evidence
2008	Convicted for possession of printing equipment restricted by law	Fined $20,000 plus court fees
2006	Convicted for possession of forged artworks	Fined $5,000 plus court fees
2004	Convicted for aiding and abetting forgery of religious artifacts	Sentenced to 6 months prison, suspended under a good behavior bond for 6 months
2002	Convicted for passing forged checks with a total value of $10,000	Sentenced to 3 months prison, suspended under a good behavior bond for 6 months

Physical Environment—Surveillance

INTERPOL Information Report—#01/2014

A meeting took place between *LETZ* and *PRESSCOT* in Dubai, UAE, on January 25, 2014. *LETZ* used a convoluted travel route to and from the meeting at the Intercontinental Hotel.

Indian Bureau of Immigration Report—#02/2014

An application lodged at the Indian embassy in Abu Dhabi by Lucien *DEWIRE* for a business visa to travel from UAE to Hyderabad, India, on February 12, 2014. *DEWIRE* has requested the visa for multiple entries over a six-month period.

Abu Dhabi Police Information Report—#03/2014

Information received from the Abu Dhabi Police advises that *DEWIRE* attended the Abu Dhabi Export Trade Office on February 1, 2014. His inquiries raised some concerns. He is reported to have asked questions about what documentation was required to enable the move of merchandise from Sudan into UAE by ship and what documentation was required to transship that cargo to the UK via truck. He wanted to take away the forms needed and asked to see an example of a completed form. When questioned, *DEWIRE* stated that he just wanted to make sure he filled in the forms correctly.

INTERPOL Information Report—#04/2014

Surveillance of *LETZ* indicated a meeting took place in the Intercontinental Hotel on January 30, 2014. The meeting was between *LETZ*,

PRESSCOT, and another European male that fit the description of *DEWIRE.*

INTERPOL Information Report—#05/2014

An influx of displaced persons has taken place in suburbs close to Port Sudan. The people are known to be staying in shanties on the outskirts of the port city. It is not known at this time why these people are attending this location.

Indian Bureau of Immigration Report—#06/2014

Lucien *DEWIRE* entered India from Abu Dhabi on February 12, 2014. He flew to Hyderabad.

Central Bureau of Investigation Report—#07/2014

DEWIRE received a male visitor at about 1900 hours on February 12, 2014, at the Hyderabad Novatel. The visitor carried a black briefcase in and out of the meeting. Surveillance could not determine what was in the briefcase. Surveillance, however, followed the male visitor after the meeting to an address in Hyderabad, and from that they identified the person as Ramesh *GUNTUR.* Surveillance established that *GUNTUR* is a 21-year-old software engineering student at the university. A check of police records showed that he had NO criminal record.

Internet Search—#08/2014

A search of public available information on the Internet (e.g., on Facebook and other social media websites) revealed that a Ramesh *GUNTUR* stated that he lived in India and is offering to sell "custom-designed" certificates and identification papers (i.e., "custom design" is considered code for fake and forged documents). His Web posting claims that his documents are of the highest standard.

Analysis

INTERPOL has requested the National Crime Analysis Agency to assist it in an investigation it is conducting into an allegation of people smuggling. The allegation is that Lucien *DEWIRE* may be implicated in this criminal enterprise.

At present the NCAA has some data that relates to *DEWIRE,* and this is in the form of his personal details, employment details, target and associates appear to practice countersurveillance during clandestine meetings, criminal record, criminal associates, and surveillance reports.

An analysis of the strengths, weaknesses, opportunities, and threats (SWOT) of these data is summarized on the next page in matrix form.[3]

Examining two factors of the SWOT analysis,[4] there are four key strengths, but eight important threats. There are several ways that police can use these strengths to "offensively" moderate the threats. The ones that present themselves include:

1. As the target's associate, *LETZ* is trained in military reconnaissance and has practiced countersurveillance while conducting clandestine meeting. The NCAA can use this knowledge when tasking surveillance units and during other data-gathering operations to ensure NCAA's involvement remains undetected.
2. NCAA's ability to provide domestic surveillance can be concentrated on the only known Indian associate, *GUNTUR*.
3. NCAA's domestic surveillance capability can be used to monitor any other associates that may become involved in the alleged criminal enterprise.
4. NCAA can construct dossiers on the target's Indian associate and the target's Indian activities from the information of INTERPOL.

Target Plan

The results of this analysis support the hypothesis that there is a criminal enterprise operating to smuggle a large number of people from Port Sudan on the Red Sea to Abu Dhabi via ship and then transport these people in containers overland via truck to the UK. To facilitate this, these findings suggest that the target, *DEWIRE*, has procured the services of a Hyderabad-based forger (*GUNTUR*) to provide the critical travel documents for the people being smuggled.

As it is not known whether the documents have been provided (e.g., electronically via e-mail) and because INTERPOL reports that there are a large number of unidentified people camped on the outskirts of Port Sudan, it is prudent to assume that a people-smuggling operation may take place soon.

If these propositions hold true, then the following recommendations would place NCAA in a good position to assist INTERPOL should the operation escalate into a full criminal investigation. In order of priority these actions include:

1. Place immigration alert on *DEWIRE, LETZ,* and *PRESSCOT.*
2. If the target or his two associates visit India, consider immediate surveillance.
3. If *LETZ* visits India, ensure all surveillance operations are aware of his propensity for violence.
4. Ensure all surveillance operatives are aware that the target and his two associates are surveillance conscious and that they practice countersurveillance.

Table 12.2.

	Pros	Cons
Internal	**Strengths** • NCAA's involvement in this operation is not known to target • Ability to provide close surveillance domestically • Can build a dossier on domestic associates quickly • Ability to monitor *GUNTUR's* Indian activities	**Weaknesses** • International locations of target and associates • Possible legal jurisdictional issues • Difficulty in conducting overseas surveillance • Reliance on INTERPOL and other agencies for intelligence on movements (e.g., ships and trucks)
External	**Opportunities** • Can work closely with INTERPOL • Access to open source intelligence on Sudan • Access to open source intelligence on associates and business associates • *DEWIRE* nonviolent "con man" • *PRESSCOT* nonviolent businessman	**Threats** • Data indicates possible people-smuggling operation from Port Sudan through Abu Dhabi via ship, then to (possible) UK via truck • Data indicates *DEWIRE* likely obtained false passports and other travel documents via Hyderabad forger (*GUNTUR*) • Large numbers of unidentified people already gathered near *LETZ's* shipping port in Sudan • Being businesspeople and with links to a crime syndicate, have potential to access large amount of money to facilitate illegal operations • Target and associates appear to practice countersurveillance during clandestine meetings • Increased involvement with other law enforcement agencies increases risk of "leaks" • *LETZ* has history of violence and is military trained • *LETZ* and *PRESSCOT* are linked to France-based crime syndicate (Mignon Syndicate)

5. Compile a dossier on *GUNTUR*, his associates, and his activities.
6. There may be enough probable cause to suspect a crime has been committed between DEWIRE and GUNTUR regarding forged travel documents—legal advice should be sought.
7. If legal advice supports a charge of false documentation regarding people smuggling, GUNTUR should be arrested and a search warrant obtained for his abode and his computer.
8. If legal advice does not support this, a search warrant for electronic intercept should be considered for his Internet connection to gather evidence regarding the likelihood of imminent supply of forged travel documents.

KEY WORDS AND PHRASES

The key words and phrases associated with this chapter are listed below. Demonstrate your understanding of each by writing a short definition or explanation in one or two sentences.

- Person of interest;
- Kipling method; and
- Target profile.

STUDY QUESTIONS

1. List the key parts of a target profile and briefly explain the types of information that would appear in each section.
2. Explain why analysts need to be conscious about the use of intelligence-related terms in their reports.

LEARNING ACTIVITY

Research an organized crime–related issue that is currently receiving public attention—this could be state/province-based or it could be national or international. For instance, issues like drugs or arms trafficking, people smuggling, money laundering, racketeering, outlaw motorcycle gangs, etc., are topical. Now, with this information, select a target profile type (e.g., target profile of an individual, problem profile, criminal business profile, gang profile, terrorist organization profile, etc.) and then create the profile. Refer to the sample target profile provided in this chapter and endnote 4 for assistance.

NOTES

1. Contrast the reports discussed here with the strategic intelligence assessment, or long-form report, discussed in chapter 19.

2. Joby Warrick, *The Triple-Agent: The al-Qaeda Mole that Infiltrated the CIA* (New York: Doubleday, 2011), 69.

3. Although many types of analysis can be used, the use of a simple SWOT analysis (i.e., strengths, weaknesses, opportunities, and threats) is often the most effective and easiest to conduct—no special training is required.

4. For brevity, only two factors were considered. Other SWOT factors could be listed in a similar fashion. For instance, there are several combinations that can be used as the basis for formulating the "Target Plan" section:

- Strengths/Opportunities—Ways that law enforcement can use strengths so that opportunities can be realized.
- Weaknesses/Opportunities—Ways law enforcement can address weaknesses in order to provide relief so that opportunities can be pursued.
- Strengths/Threats—Ways law enforcement can use strengths to "offensively" moderate the threats.
- Weaknesses/Threats—Are there defensive actions that law enforcement can implement that will protect against threats?

13

♨

Tactical Assessments

The topic discussed in this chapter follows on from the previous chapter on operational-orientated intelligence reporting by looking at another type of report, including:

1. Introduction;
2. Tactical versus strategic assessments;
3. Tactical or operational assessments;
4. Key parts of a tactical assessment;
5. General considerations for tactical and operational reports; and
6. An example of a tactical assessment.

INTRODUCTION

A tactical assessment is a type of intelligence report that takes a wider view of a situation than does the target profile. Although not in the league of a strategic assessment, the objective of this type of report is to shift the emphasis away from being reactive to being anticipatory. Whereas the target profile focuses on a particular individual (or company or organization), the tactical assessment looks at the problem in a broader way. The audience for a tactical assessment would be a whole-of-agency group that deals with tasking or an interagency group responsible for coordinating assets across several organizations (which could include agencies at different levels of government, or sectors—e.g., military and law enforcement).

For example, if a target profile examines Jack Knife, who is involved in drug trafficking, then a tactical assessment may not only look at where Knife fits into this picture but the extent to which the drugs are trafficked, those involved (buying and selling), the social and economic impact on

the community from the direct effects of the illicit drugs, and the indirect
and consequential effects of the ill-gotten profits on, say, corrupting legitimate business, political, or regulatory officials.

TACTICAL VERSUS STRATEGIC ASSESSMENTS

Although a tactical assessment has the hallmarks of a strategic assessment that will be discussed in chapter 19, it falls short of its companion
because it focuses on short-term objectives that would result from immediate action to prevent further illegal or otherwise unwanted activity. A
strategic assessment looks at a longer time frame and usually features
several recommendations that in combination need to be put in place in
order to defend against the threat, or neutralize it, or treat a risk.

Agencies will have their own house style for this type of report (and it
may even be known under a variation of this name), but it will follow this
template to a large degree. Analysts may optimize the sections and format to suit the intelligence project they are working on. Because the life
span of a tactical assessment is short, a new or revised report may have to
be required to be produced weekly or monthly, whereas a strategic
assessment is a unique report on a specific topic.

TACTICAL OR OPERATIONAL ASSESSMENTS

Although these analytic reports are termed *tactical assessments*, they might
be better referred to as *operational assessments* if the terminology discussed
in chapter 2 under the heading of "Taxonomy of Intelligence Research" is
applied in the strict sense. By definition, if the issue under investigation
is broader than what is happening at the tactical level (e.g., investigation
and apprehension), but not predictive as strategic intelligence would be
(i.e., long-term implications), then it is operational. However, convention
has been to use the term *tactical*, so that is what has been used here.[1] This
anomaly is worth noting, as it may cause confusion when developing an
assessment—it would be easy to mistake the scope of the report based on
this loose use of the term.

So, like target profiles (discussed in chapter 12), other names could be
applied to tactical assessment depending on the issues under investigation. Take, for instance, these examples—*criminal business assessment, terrorist organization assessment*, or a name relating to a specific crime
problem, like a *stolen vehicle assessment*. Variations and adaptation can be

"The value of tactical [assessments] to law enforcement administrators beyond the immediate arrest and prosecution can be extremely important in determining how, where, and with what degree of intensity resources should be allocated."[1]

1. Justin J. Dintino and Frederick T. Martens, *Police Intelligence Systems in Crime Control* (Springfield, IL: Charles C Thomas, 1983), 114.

applied to suit the topic being studied. But whatever is used, it needs to be clear what type of report it is and that it is different from a target profile or a strategic assessment, which are at the other two ends of the report continuum.

KEY PARTS OF A TACTICAL ASSESSMENT

Introduction

This section provides brief details about the legal basis or agency policy which gives the analyst authority to compile the report, and the name of the authorizing officer (e.g., the intelligence unit's manager), date, and file number for audit purposes.

Background

The report's introduction section will contain a statement of the assessment's aim or objective as well as a description of how the problem or issue under investigation arose. Like a target profile, this section needs to provide an acknowledgment of the legal basis or agency policy that provides the authority for compiling the report and the name of the authorizing officer (e.g., the chair of the tasking group or the officer in charge of the interagency committee).

Aim

The aim is in essence the report's research question. It acts to guide the research and keep the inquiry focused on a specific issue. This helps keep the investigation from "wandering" into areas that could be interesting but not relevant to the outcome of the issues being perused.

Current Situation

This section is comparable to the various descriptive sections that appear after the background section but before the analysis section of a target profile. In fact, this section may comprise several subsections of descriptive data about the problem. These data can come from other intelligence reports (e.g., target profiles); open sources (e.g., Internet); academic studies (PhD dissertations and master's theses); or articles in scholarly journals, books, or government publications (e.g., bureau of statistics), to cite a few.

In addition to describing the phenomenon, the analyst can explain the ramifications of the problem in its historical, social, economic, political, religious, cultural, or anthropological context and its extent in the jurisdiction (e.g., the region, the nation, or around the globe). But it is important not to interpret the information here in this section—that comes after analysis, in the prognosis section. The analyst could explain any progress being made, pitfalls encountered during the investigation, or the implementation of interventions to date. This information sets the scene for the next section, which is the analysis.

Analysis

As in the analytic section of the target profile discussed above, this is where analysts present the results of their analysis using techniques such as statistical analysis, network analysis, force field analysis, SWOT, PESTO, or others depending on the scope of the research question.

Prognosis

The prognosis section is essentially a discussion of the analysis but extrapolates from the findings to assess what is revealed about the activity under investigation and what can be done to provide relief. The analyst may consider a change in focus from what is or was being done to a modified or totally new approach or a shift in priorities using the same interventions, and so forth.

It is common to present this discussion within the frame of the results of the analysis. Although framed in the logical order of the analysis, this is an exercise in deductive reasoning in a narrative form—or "thinking out loud" about each of the issues discovered in the analytic process that preceded it.

In order to produce a range of options for decision makers to consider, this section talks about the agency's strategic mission/goals or key performance indicators (KPIs) and how possible interventions may impact these benchmarks. Decision makers around the conference table can then argue priorities and resources according to "what works," "best value," or "best practice." Issues that might be discussed in the prognosis section could be generated from any one (or more) of the topics contained in the *five Is model*:

- **Intelligence.** Issues relating to information gathering, collation, and analyzing (past or future);
- **Intervention.** Tactics to block, disrupt, weaken, or eliminate "the problem";
- **Implementation.** Translating the goal of the proposed intervention (theory or principles) into practical methods in the field;
- **Involvement.** Ways to get other agencies (or companies, organizations, and individuals) to contribute somehow to being part of the implementation of the intervention(s); and
- **Impact.** How the problem will be evaluated and by whom. The evaluation may be simple or complex, but as the problem is one of a tactical nature, a basic evaluation is most likely all that is needed (i.e., an output-based evaluation rather than one that is outcome focused).

Recommendations

Intelligence managers dealing with tactical issues that are within the scope of this type of assessment will require options for consideration. Stemming from the previous section, the analyst needs only restate the range of options available. This can appear in the form of a bulleted list to simplify what is possible. If there is a preferred option, this can be highlighted in some way—for instance, appearing first in the list with the other options appearing in a list below in diminishing order, with the least preferred at the end.

To help frame a set of recommendations, use the straw man technique discussed in chapter 20. This technique allows decision makers to understand the strengths/benefits of the preferred option when contrasted with other options.

Appendixes

Because a tactical assessment is a focused report like the target profile, the attachments that may be included need to be kept to a minimum. Examples might include a map, a photograph, or an organizational chart showing complex relationships that can be more helpful than a drawn-out narrative.

GENERAL CONSIDERATIONS FOR TACTICAL
AND OPERATIONAL REPORTS

When writing the intelligence report, avoid placing facts, inferences, rec-ommendations, and analysis within the same sections. Consider using a *funnel approach* to writing the report. That is, start from the general, work to the specific—like the shape of a funnel.

Each section needs to help explain the "story" so the narrative flows logically. If a section of the report discusses the current situation, it could contain the facts that are known. The analysis section that follows takes these facts and subjects them to one or more analytic techniques so that insight can be developed. This analysis should also place some level of likelihood/probability on these scenarios (or hypotheses) with discussion about the limits (e.g., based on, say, a risk or inferential analysis). This narrative would then lead the reader to the plan/recommendations sec-tion—like a funnel. Or simply, it's a story that has a beginning, middle, and an end.

Unlike fictional stories, the beginning of an intelligence story contains facts—what, why, when, how, where, and who (known as the Kipling method). The middle tends to contain analysis, and the end contains the report's conclusions, recommendations for action, or policy options, all of which can appear in a variety of forms (often prescribed by the employ-ing agency in a standardized template).

Analysts should resist the temptation to append their analytic results at the end of the report, as it could mean they failed to refer to them in the narrative (i.e., to help explain the story). Alternatively, referring the reader to peruse the appendices is getting the reader to do the job of the analyst.

In the same vein, tables of data or matrices should not be simply dropped into the analysis section of the report, as a lot of detail is often contained in these. It is better to summarize the key aspects of these anal-yses, and in case the reader wants to see the big picture or how you arrived at your conclusion, place the diagram or other figure/table in an appendix.

The reason for suggesting this writing approach is that the key points become the basis on which the analyst will make recommendations. Pres-enting the assessment's recommendations as a set of options is an impor-tant aspect of the report. Couching the options in terms of a threat assessment, a risk assessment, a budget, or the human resources available are excellent approaches. To the reader, such features make the recom-mendations actionable.

These guidelines are not hard-and-fast, but following them will increase the likelihood that the message contained in an intelligence report will be understood by decision makers and, if so, acted upon.

AN EXAMPLE OF A TACTICAL ASSESSMENT
Introduction

Date:	April 1, 2014
Authorizing Officer:	Director, National Crime Analysis Agency (NCAA)
Crime Analysts:	International Organized Crime Group (Delta X-Ray)
In the matter of:	Section 50, Anti-People Smuggling Act, 1992
File:	I-0038/2014

Background

Six weeks ago the International Organized Crime Group (Delta X-Ray) was tasked to develop a target profile on Lucien *DEWIRE* in relation to an allegation that he was involved in a criminal enterprise to smuggle people. This criminal enterprise was alleged to have been operating in France, the United Kingdom (UK), the United Arab Emirates (UAE), and the Republic of India. The results of the target profile analysis supported the hypothesis that there is a criminal enterprise in operation that is smuggling large numbers of people from Port Sudan on the Red Sea to Abu Dhabi via ship, and then transporting these people in cargo containers overland via truck to the UK. The report's findings suggested that *DEWIRE* procures the services of a Hyderabad-based forger (*GUNTUR*) to provide the critical travel documents for the people being smuggled.

Aim

The aim of this tactical assessment is to assess whether the matter involving *DEWIRE* is an isolated case of people smuggling or whether there are other criminal enterprises operating in this illicit industry.

Current Situation[2]

People smuggling is the illegal transportation of people across international borders. This happens either in secret (covert) or clandestinely (openly under deception). It usually requires some false travel and/or identity documents and involves a person or group that facilitates the operation. The facilitator is paid for their services, which include the obtaining of the false documentation.

It is important to note that people smuggling differs from *people trafficking*, which has been described as modern day slavery. People trafficking is characterized by force and coercion, sometimes coupled with deception, where the subject people are deprived of freedom and placed in a position of debt bondage. The most common forms of people trafficking involve sex exploitation and labor exploitation. People trafficking is not the subject of this assessment.

Research shows that there is no one single reason why people consent to being smuggled across borders, but the chief reasons include new social or employment opportunities or reuniting with family in the destination country.

The fees paid to people smugglers can be many times the cost of legitimate travel, and there is no guarantee that the smuggler will fulfill their promise to carry out the smuggling operation, or if they do, there is no assurance the operation will be successful. Anecdotal evidence shows that people pay smugglers between $4,000 for a wholly land-based infiltration to $75,000 for an operation involving sea travel. Case studies show that seaborne attempts at people smuggling end in capsized boats and sometimes mass drowning. The rescue and hospitalization of people overboard (see figure 13.1) and the subsequent investigations into the incidents are costly and time consuming for authorities. Land-based

Figure 13.1.   Rescue of a woman overboard by Hong Kong Police.
Photograph by author

smuggling cases show that people have been killed through asphyxiation and heat, again, consuming large amounts of public resources to deal with the aftermath of the tragedy.

People who contract the smuggler's services can be subject to physical and/or sexual violence *en route*. These people can also be subject to blackmail by the smugglers once they enter the destination country for fear of exposure of their illegal residence status, though this is not universal and appears to be in a small number of cases.

The current case involving *DEWIRE* has all the hallmarks of a classic people-smuggling operation. The target profile compiled by the Delta X-Ray analytic team of the International Organized Crime Group supports such a hypothesis, and its report provides compelling evidence to back this proposition. Since the target profile was developed, INTERPOL has communicated its intent to conduct a criminal investigation into the matter.

Because people-smuggling operations are conducted in secret, it is difficult to know the exact scale or how many criminal enterprises are involved. However, researchers studying this phenomenon estimate that as many as 800,000 people may have entered the European Union in the last calendar year. This estimate is based on a United Nations methodology that highlights its tenuous conclusion but considers it a reasonable approximation. This indicates a very low risk of capture, and official statistics support this.

Studies show that nearly all countries are affected by people smuggling in terms of being a country of origin or a destination country. This has the effect of increasing the chances of crime and corruption in countries touched by the smuggling operations because of the large profits that are generated from the enterprise.

Analysis[3]

The purpose of this tactical assessment was to assess whether the *DEWIRE* matter is an isolated event or whether this activity is likely to involve other criminal enterprises. When the facts presented above were weighted in a pros-and-cons analysis (see table 13.1), there was an overwhelming amount of evidence to support the conclusion that the *DEWIRE* matter is not isolated.

It could be concluded that there is a strong probability that other smugglers are operating in or through the jurisdiction overseen by the National Crime Analysis Agency.

It could also be concluded that given the geographic location this jurisdiction plays in a seaborne transportation route, the amount of money involved is at the higher end of the UN's estimate, and this in turn

Table 13.1.   Summary of Pros and Cons

Pros	Cons
• Smuggling is performed covertly and clandestinely	• Disproportional cost when compared to legal travel
• Participants are willing subjects	• Dangerous
• The rewards are very high	• Possibility of being caught (low)
• High likelihood of success (low risk of being caught)	• Possibility of being hurt or exploited (low)
• Enforcement agencies cannot cover all the ports and border crossings effectively due to the numbers involved	
• Payment to corrupt officers lessens the likelihood of detecting smuggling operations	

increases the likelihood of fueling graft and corruption within the jurisdiction. For instance if the UN's estimate of 800,000 is reflective of the magnitude of the problem, and a conservative estimate for the cost is calculated at $10,000 per person, then this equates to an $8 billion per year industry.

Prognosis

Although the director of the National Crime Analysis Agency has authorized the NCAA to join the task force probing the *DEWIRE* allegation, the priority question for the agency is whether the current matter is isolated or is the issue wider spread. This assessment concludes that the problem is more than likely to be far greater than a single criminal enterprise, and given the potential scale of this illicit industry, it is likely to be an ongoing problem that will have implications for other crimes faced by the community.

Recommendations

At this juncture it is important to understand the extent of the people-smuggling business within the jurisdiction. This is an intelligence issue, and as such the Strategic Analytic Team (Delta Bravo) of the International Organized Crime Group should be tasked to develop a strategic assessment.

The strategic assessment should describe the role the jurisdiction plays in the smuggling process—source, destination, transshipping, facilitating, or other. It should also include an estimate of the size of the problem in terms of number of crime enterprises, the volume of people being trafficked, as well as the ramifications in terms of secondary crime or consequential crime within the jurisdiction. Finally, the assessment should provide a range of options for addressing each of the issues raised in the report.

KEY WORDS AND PHRASES

The key words and phrases associated with this chapter are listed below. Demonstrate your understanding of each by writing a short definition or explanation in one or two sentences.

- Five Is model;
- Funnel writing approach; and
- Tactical assessment.

STUDY QUESTIONS

1. List the key parts of a tactical assessment and briefly explain the types of information that would appear in each section.
2. Explain why using a "funnel" approach to writing reports is a useful method.

LEARNING ACTIVITY

Using the target profile developed in chapter 12's learning activity, continue your research in the issue you selected. Using that target profile and the additional information you have discovered on your topic, create a tactical assessment. Refer to the sample tactical assessment provided in this chapter and endnotes 2 and 3 for assistance.

NOTES

1. See, for instance, the United Kingdom's *National Intelligence Model* (London: National Criminal Intelligence Service, 2000).
2. This section of the sample report has been abbreviated in order to demonstrate the type of information that could be presented and the manner in which it should be written. If this was an actual case, this section might be two or three

times the length it is and the narrative would be supported by actual references to the scholarly literature that supports the facts or assertions made. With regard to in-text referencing, it is suggested that the author-date (Harvard) style of referencing is used. This is one of the easiest to use and produces an uncluttered appearance in the text. This has the effect of not slowing down the reader, yet offers substantiation to claims and assertions made. There are many free guides that explain the Harvard method of referencing on the Internet.

3. Like the abbreviated "current situation" section of this sample report, the analysis section may be longer and contain a more detailed summary of the analysis. But for the purposes of demonstrating how such a section might look, the information provided here is sufficient.

14

§

Vehicle Route Analysis

This topic will discuss two important considerations that underpin the establishment of safe travel for friendly forces in war zones and clients in hostile urban areas:

1. The planning framework; and
2. The analytical process.

INTRODUCTION

Although the concept of vehicle route analysis will be discussed here in terms of military or paramilitary operations, the concepts apply equally to those operating in other security environments. For instance, an analyst could carry out such an analysis in regard to, say, a threat of kidnapping or vehicle hijacking.

It is an invaluable way of thinking for those who recognize that preparation is the best form of defense, and in relation to forces operating in zones that are engaged in the Global War on Terror, it is vital. Nevertheless, for personnel who are operating in equivalent zones of dispute—say, urban areas of industrialized countries where gang violence is a hallmark—vehicle route analysis will be important. It is not only a structured way of thinking for intelligence analysts, but a practical skill for security operatives.

Although this analytic method will have primacy with military audiences, it will also be important to security personnel who work in the area of close personal protection. Moreover, it will be vital to intelligence officers engaged in the planning of vehicle or convoy travel through hostile areas. Arguably, this is because vehicle route analysis is at the heart of all vehicle movement—whether it is a single one-off passenger "pick-up and drop-off" or a multivehicle motorcade carrying VIPs to various meetings over a number of days.

THE PLANNING FRAMEWORK

The planning framework for route analysis is based on ten interrelated plans. These plans in turn form a stepwise approach to how an analyst or operative conducts such an overall analysis.[1] Step by step, an analyst conducts:

1. a map reconnaissance;
2. a hazard analysis;
3. a crisis response planning;
4. a travel surface analysis;
5. an attack site analysis;
6. an attack type analysis;
7. a vehicle hardening plan;
8. a countersurveillance plan;
9. a communications plan; and
10. considers supplementary information.

This framework needs to be placed into a geography-independent context—that is, these steps apply regardless of whether the vehicle under protection is traveling through one of the world's major cities or through a war zone. But the geography context will be different—vehicles carrying VIPs through, say, Amherst, Massachusetts, will be different from the same convoy traveling through Hyderabad, India.

THE ANALYTICAL PROCESS

Map Reconnaissance

So how does this analytic process work? First, map reconnaissance is conducted, which enables the analyst to determine several alternative routes to and from the destination. Having at least three routes introduces an unpredictable variable to foil any hostile threat. This analysis identifies features such as choke points, major thoroughfares, and potential safe havens should an attack take place. These data can be gleaned from a variety of maps—publicly available street directories, government-produced topographical maps, commercial or classified satellite photographs, and

Figure 14.1. Hyderabad, as well as other large busy cities around the world, poses challenges for vehicle route analysts.

Photograph by author

maps used by utility companies (yes, even Internet-based mapping facilities can be used).

This approach is encouraged, as it combines data to gain a greater understanding of the route's total features. Ideally, if some form of geo-spatial mapping system is available, this would be the analyst's best method. If not, the time-honored overlay mapping system discussed in chapter 15 will perform more than adequately (whether this is via flipcharts or computer projection software/electronic presentation).

Hazard Analysis

Then, hazard analysis identifies any areas along the travel route that might pose a direct danger by, say, placing a restriction on a vehicle's movement or providing cover for a person or group who may threaten an attack (i.e., *threat agents*, as discussed in chapter 18). It also identifies any areas along the route that might present an indirect hazard by providing a support to the danger posed, thus allowing a threat agent to use an open area for an attack. For instance, a restriction could be as simple as a traffic light, stop sign, or highway underpass—anything that gives a potential threat agent the ability to control the area as the vehicles pass.

Examples of indirect hazards could be a highly built-up urban development along an otherwise clear stretch of roadway. Although the roadway would not be a good place to launch an attack, the attackers could launch a hit-and-run attack and then quickly disappear into the adjacent "urban jungle," thus avoiding identification, capture, or engagement with the vehicle security detail. Travel hazards are any feature that either impedes a vehicle's ability to change course or imposes control over a vehicle, restricting its movement to a course that makes it more vulnerable.

Crisis Response Plan

Having done this, the next step is to develop a crisis response plan for each choke point or route hazard. It is a predetermined set of guidelines that provides the driver with what to do if attacked at each vulnerable point. It provides options for the best direction for exiting, how far the next safe haven is, whom to call for assistance, and most importantly, which roads not to take.

In planning for calls for assistance, the proposal should allow for handheld radio transceivers to be carried in the vehicle. Should the occupants have to abandon the vehicle and prepare for escape and evasion on foot, a handheld radio will be invaluable for providing details of the situation and coordinating a rescue (see communications plan below).

Scheduled halts need to be factored into these plans if the vehicle or convoy will be traveling for several hours. A stop of about fifteen minutes after the first hour is required to help maintain concentration and alertness. Thereafter, a stop of about ten minutes every two hours is required. When these stops are planned, location, cover, rapid egress, and provisions for human comforts need to be considered in regard to the overall safety concerns.

In determining the travel hazard options and crisis response plan, an analysis also needs to be conducted of the surfaces the vehicles will be traveling over. This analysis will need to take into account conditions for various seasons (if the passengers will be transported over a long period of time) or for the different conditions that are likely during the season the route is used (for shorter-term taskings).

Travel Surface Analysis

Considering road surface options is an important part of the overall analytic process, because driving conditions differ for the same road surface. For instance, an otherwise perfectly passable dirt road in the dry season could present a very difficult hazard if only a few centimeters of rain fell by turning the top part of the road into a slippery sheet of mud, slowing vehicles or causing them to lose a substantial measure of maneuverability. This road condition could create a perfect killing zone for attackers.

Having analyzed several alternative routes, the safe havens associated with each, route hazards, and road surfaces features, the next aspect of the overall route plan is to analyze where each attack is likely. This is a more specific type of analysis based on the results of the other contributory analyses. It produces a set of descriptive narratives (or diagrammatic figures) that identify the most probable kill zones—where on the route an attack might start and finish, and the areas likely to conceal the threat.

Attack Site Analysis

The final part of the analysis is an examination of the type of attack likely. This is based on past experience. Research shows that attackers tend to use methods that they are at ease with; hence it is likely that the threat will employ weaponry and tactics they have used before. For example, there are groups that have used improvised explosive devices, while others prefer stand-off attacks using sniper weapons or rocket-propelled grenades.

There are many possible types of attacks, from roadblocks to set up a hit-and-run ambush, to assaulting an entire convoy in one kill zone. It is important to canvass all possibilities using one of the idea generation techniques discussed in chapter 6 (i.e., various data collation techniques) and then assess the results using a technique like competing hypothesis (see chapter 11).

Attack Type Analysis

The analyst needs to research the groups that operate in the areas the vehicle or convoy will travel and apply these data in the context of the route's vulnerable points and surface conditions. Whether this process is via scenario building or the analyst's agency's preferred method of theorizing, is not critical. What is important is to acknowledge that there are several possible attack variants, and each should be considered—urban static attacks, urban moving attacks, rural static, rural moving, and attack sites for stand-off weapons.

In conducting this type of analysis, it is important to distinguish whether the threat agents are insurgents or guerrillas, or if they are regular military forces. This is because the tactics and the arms used may be vastly different. Take for instance an attack by regular military forces—this might include ambush via air using helicopter gunships or artillery, whereas a group of local insurgents may use an improvised explosive device, a rocket-propelled grenade launcher, or sniper fire using small arms.

Vehicle Hardening Plan

If attack issues are found to present a danger to the vehicle or convoy through one of the proceeding analyses, then the vehicle or convoy may require hardening to mitigate the risks. As there are as many methods of adding protection to vehicles as there are vehicle types, the specifics will not be discussed here. Nevertheless, it is adequate to say that an automotive engineer or mechanic should be consulted to determine the type of armament that can be installed to afford the level of protection required.

Adding protective shielding is not a simple matter, as the vehicle's capabilities need to be factored into the fortified design. Factors such as the vehicle's weight-carrying ability and its steering, accelerating, and braking performance need to be carefully considered.[2]

Countersurveillance Planning

To draw all this thinking together is to assess the advantages of using countersurveillance operative(s) at fixed location(s) along the route. This countermeasure is contingent upon the analyst's assessment of not only the overall travel risk but a realistic assessment of how much benefit countersurveillance will have.

Remember that deploying covert field operatives may place these personnel in harm's way, creating an even more dangerous situation—they may be kidnapped, tortured, or killed. Clearly this decision would need to be done in consultation with the leaders of the countersurveillance team and the vehicle security detail.

Analysts should also consider camouflage, concealment, and deception as a means of foiling surveillance. Camouflage, of course, is designed to help the vehicle blend into the surroundings. Traditionally, this is thought of as snow, woodland, jungle, or desert camouflage. However, if the vehicle or convoy is operating in the urban area of a foreign city, it may mean making the vehicle look similar to those of the local inhabitants that are traversing the city's streets.

Concealment can be as simple as covering the load of a vehicle with a tarpaulin to conceal the goods being carried. And deception can be any tactic that diverts the attention of a threat agent away from the vehicle or convoy—for example, a dummy convoy that leaves minutes early so the attacker's surveillance team sends the assailants in a different direction.[3]

Communications Plan

Providing a situation report (sitrep) on the progress of the vehicle or convoy is fundamental. Usually these reports go to some form of operations center. Therefore, two-way radio equipment needs to be robust and operatable. It needs to be manufactured to military specifications—it cannot include the off-the-shelf commercial-quality radios that are sold through retail outlets. This is because "mil-spec" (or "mil-std") equipment is designed to tolerate dust, heat, cold, and vibration to much higher levels than common commercial-grade radios. Mil-spec radios are sold on the commercial market but are advertised as "public safety" radios and marketed to police, fire, and emergency services.

The type of radio needs to be determined in this plan too—UHF[4] radios do not have the range of HF[5] sets. Such a consideration is a critical factor if the vehicle or convoy is traveling several hundreds of kilometers as opposed to traveling ten kilometers to the next township. UHF radios may also suffer in deep ravines, during thick sandstorms, or if there is dense smoke, whereas VHF[6] radios may perform somewhat better in these conditions. A radio engineer needs to be consulted to help determine the most suitable equipment as well as the frequencies for the operation. It might be determined that a transceiver for each band (i.e., HF, VHF, and UHF) should be installed in the vehicles. Procedures for each band and frequency can then be selected for either the entire mission, parts of the mission, or in emergency situations.

Once equipment requirements have been determined, a set of radio call signs needs to be established, with check-in times and frequencies specified. Moreover, code words for emergency and routine events need to be understood, and all radio operators need to be briefed on the times/areas where radio silence is to be maintained, if required.

If vehicles are traveling in a convoy, then the communications plan needs to specify the frequencies and procedures for intraconvoy communications, perhaps using UHF frequencies, as these lessen the chance that these signals can be intercepted at distance, as discussed in endnote 4. These radios will usually be secondary sets mounted in the vehicle and operating on different frequencies than that being used for communications back to the operations center (i.e., on HF[7] and/or VHF).

Handheld radios also need to be incorporated into the plan as part of the crisis response plan. If personnel need to evacuate the vehicles in an emergency, then the vehicles' mounted radios will be of no value for providing details of the emergency, location status (locstat), or situation reports.[8]

Supplements

While developing the various analyses that contribute to the overall vehicle route analysis, analysts are likely to consider it necessary to append material that does not fit neatly into the narrative of the body of the report. Maps, diagrams, and aerial photographs immediately come to mind, but there could be other material too: for instance, photographs of safe houses, threat agents, route hazards, and so on. If there is such material, these pieces of information can be appended to the end of the report and referred to in the appropriate section. This type of information can help aid understanding and provide reference in case of emergency—an example could be a list of emergency frequencies.

SUMMARY

Although we have only examined the analytic considerations of route analysis here, a more thorough understanding of other aspects of driving in high-risk areas would round off this discussion—whether it is driving in Baghdad, Beirut, Bogotá, or Bondi Beach. The technical content of that discussion is likely to be based on real-life experiences. Therefore, if analysts and operatives alike can gain an understanding of the allied issues of route analysis, it would greatly assist them in performing the various subanalyses just outlined—for instance:

- vehicle dynamics;
- evasion maneuvers;
- pros and cons of vehicle armoring;
- motorcade defensive tactics;
- effecting an *en route* rescue;
- vehicle safety equipment; and
- how to conduct vehicle bomb searches.

Although the basics of how to conduct a route analysis were discussed, doing so assumes a good deal of subject knowledge that is beyond the scope of this book. Nonetheless, understanding the theoretical framework suggests to those who are interested in performing this type of planning that they should acquire knowledge about such things as the history of where VIPs have been ambushed using explosives or small arms; where they have been kidnapped *en route*; or where they have been overrun during the attack, and so on.

Also, the interested analyst or operative would benefit from seeking out books that discuss vehicle dynamics, evasion maneuvers, and armoring of a vehicle.[9] A case study approach is well suited to this type of research, as it will highlight the countermeasures potentially available and how to apply them in various cases by understanding the strengths and weaknesses of both attackers and victims.

KEY WORDS AND PHRASES

The key words and phrases associated with this chapter are listed below. Demonstrate your understanding of each by writing a short definition or explanation in one or two sentences.

- Attack site;
- Attack type;
- Countersurveillance.
- Crisis response;
- Hazard;
- Reconnaissance; and
- Travel surface.

STUDY QUESTIONS

1. Outline the seven steps for conducting a vehicle route analysis.
2. List three potential sources of geospatial information for a map reconnaissance.
3. Discuss why it is important to conduct a travel surface analysis.

LEARNING ACTIVITY

Suppose you have been asked to help construct a plan for a visiting VIP. The person will travel from the airport nearest to where you work and go to a meeting in a covert location five kilometers away. Using a map from one of the sources you identified in the second study question above, draw a circle around the airport with a five-kilometer radius. Now select a populated street somewhere along that radius line, and using the seven steps of vehicle analysis, create a plan. The plan need not be immensely detailed. What you are trying to achieve is an understanding of each of the elements that comprise the plan. Once compiled, reflect on these issues: were you able to populate each subanalysis to your satisfaction? If not, what was lacking—subject knowledge? More information about the area? Where might you find those details if you were to conduct this exercise as a real-world event? What subject knowledge might you want to feel confident in completing this type of analysis if it were not an exercise?

NOTES

1. See, for instance, Robert H. Deatherage Jr., *Survival Driving: Staying Alive on the World's Most Dangerous Roads* (Boulder, CO: Paladin Press, 2006).
2. As an example, see U.S. Department of the Army, *FMI 3-07.22, Counterinsurgency Operations* (Washington, DC: U.S. Department of the Army, 2006).
3. For more on countersurveillance, see Hank Prunckun, *Counterintelligence Theory and Practice* (Lanham, MD: Rowman & Littlefield, 2012).
4. UHF is the abbreviation for *ultra-high frequency*. Frequencies in this part of the radio spectrum range from 300MHz to 3GHz (i.e., 3,000MHz). They are line-of-sight (i.e., point-to-point) and excellent if the signal needs to be "contained" so that it does not "stray" beyond the operational area. This could be important if the vehicles' signals may be monitored by third parties at some distance. If the monitoring station is not within line-of-sight, it reduces the chance of interception. Communications security can also be enhanced by the use of voice encryption. Even commercial-grade encryption units provide a very high level of security, though it may not be to the classification of Secret or Top Secret. Nevertheless, if communications security is required for a short period of time (i.e., a

matter of hours or a day), then any intercepted encrypted signals may not be able to be unencrypted in that time frame, and if they are eventually unencrypted, the operation would be over and the convoy safely at its destination. A radio engineer needs to be consulted about the technical specifications of such equipment and what it can offer the mission in terms of communications security. Signals in this band are affected by variables involving atmospheric conditions at ground level and terrain. However, because UHF operates on line-of-sight, it is ideally suited for communications through satellites, as these signals pass directly through the ionized layers of the upper atmosphere. If a communication satellite is available, UHF equipment may be suitable for intraconvoy communication (i.e., using a simplex frequency) as well as long-range communication (i.e., using duplex frequencies). See Martin Davidoff, *The Radio Amateur's Satellite Handbook* (Newington, CT: American Radio Relay League, 1997).

5. HF, or *high frequency*, spans from 3MHz to 30MHz, and it is a band that is capable of delivering reliable communications that range from several kilometers to transcontinental distances. It is also known as the *shortwave* band. But because of the great distances these signals can travel, interception by third parties increases. See, for instance, Harry L. Helms, *How to Tune the Secret Shortwave Spectrum* (Blue Ridge Summit, PA: Tab Books, 1981), and Oliver P. Ferrell, *Confidential Frequency List*, fifth edition (Park Ridge, NJ: Gilfer Associates, 1982). Intelligence agencies continually monitor these bands, so if the convoy's radio traffic is sensitive in any way, one should assume some intelligence agency somewhere in the world may be monitoring (James Bamford, *The Puzzle Palace* [Boston: Houghton Mifflin Company, 1982]).

6. VHF refers to *very high frequency*. This band encompasses frequencies between 30MHz and 300MHz. It lies in between the HF band and the UHF band and is a useful band, as it is not as susceptible to atmospheric conditions and ground terrain as are signals in the UHF band.

7. Because of the nature of HF communications, it requires a large antenna. For instance, at 27MHz a quarter-wave antenna is 9 feet (2.6 meters). At 14MHz, this would be approximately 16 feet (5 meters). And at 5MHz, a vehicle-mounted quarter-wave antenna would be about 47 feet long (14 meters). Granted, HF antennas are usually "shortened" by the addition of loading coils and tuned via automatic tuning devices built into the radio sets, but the antennas are still very large. Their size makes the vehicle stand out and can act as a beacon calling attention to it and the convoy it is in.

8. The basics of emergency communications are covered in such texts as American Radio Relay League, *Emergency Communication Handbook*, Steve Ford, ed. (Newington, CT: ARRL, 2005); and Michael Chesbro, *Communications for Survival and Self-Reliance* (Boulder, CO: Paladin Press, 2003).

9. See, for example, Ronald George Ericksen II, *Getaway: Driving Techniques for Escape and Evasion* (Port Townsend, WA: Loompanics Unlimited, 1983). This is an excellent slim-volume paperback that focuses on practical skills and realistic vehicle modifications.

15

၍

Geographic Analysis

This topic focuses on three elementary ways of presenting spatial data:

1. Maps;
2. Overlays; and
3. Mosaics.

MAPS

Basics

A map is simply a two-dimensional graphic representation of part of the earth's surface. It is drawn to a scale (except for those that are created in the field and termed *mud maps*) by cartographers and is always oriented with a bearing to north. Mapmakers incorporate certain devices into the map in order to make reading easy—universal symbols for various features in the physical world as well as colors and lines to augment the representations.

While some maps are three-dimensional—being constructed out of sand, clay, wood, plastic, paper, or cardboard—our discussion here will limit itself to two-dimensional, paper-based maps and digital maps. These can be produced by even the most modest forms of spatial analyses, and the results can easily be displayed for decision makers' consideration.

Maps are important, as they allow an analyst to simultaneously display a number of pieces of information that may prove confusing in narrative form—recall the adage *a picture is worth a thousand words*. Further, through the medium of a map, analysts can conduct their analysis-in-chief—say, comparing distances and lines of sight and evaluating access, travel routes, and so on. Because each map contains numerous data items, it is a self-contained database that shows the existence and location of many ground features as well as the distance between them.

Arguably, the most frequent user of maps in intelligence work is the military. Although some use is made by law enforcement agencies with the mapping of crimes and crime hot spots, their use is not universal. If a map is to be incorporated into an intelligence report, it is imperative that the analyst knows the cartographic skill level of the reader. If, for instance, the reader has no knowledge beyond reading a city street map, then this is the maximum level of sophistication that should be presented in the map. More detail may cause the reader to lose interest, and therefore, the analyst will lose his or her audience.

Sources

Maps can be sourced from many places—from government departments that are responsible for land surveying to commercial companies that specialize in the production of high-quality maps. At the time of this writing, the Internet featured several digital sources of maps, including maps that show satellite images of the earth. The resolution of these images is good, and the user can select the level of magnification required. These satellite images can also be displayed as a simple map with terrain features or in composite form (i.e., terrain and image). Other Internet-based mapping systems available at the time of this writing included those that showed street-level maps (views) of cities around the world and a virtual view of the oceans' landscape.

Through the use of desktop computer workstations and software packages, analysts can convert paper-based maps to digital images for projection in oral briefings, incorporate them into digitally created documents, or save them to a secure intranet site within the host agency (e.g., in some form of electronic knowledge base as part of the agency's *basic intelligence* holdings).

Aerial Photographs

Aerial photographs are photographs taken from any of a variety of airborne platforms, such as fixed and rotatable wing aircraft, unmanned drones, hot air and gas-filled balloons, and satellites. The aerial photograph can be used as either a supplement to a cartographic map or as a map substitute.

Why would an analyst consider displaying intelligence in the form of aerial photography? On the one hand, a topographic map may have been created some years previously and, hence, could be obsolete. On the other, an aerial photograph taken recently (via the analyst's information collection plan at the start of the research) shows any changes in landscape as they occur.

In this regard, a display of aerial photographs and maps provides more information than either alone. Photographs can be obtained readily, whereas maps can take weeks to prepare. Changes can be noted and compared daily or even hourly if the issue is that important (e.g., the Cuban Missile Crisis of October 1962). The downside, however, is that aerial photographic interpretation is both an art and a science that takes considerable skill and training (as an illustration, see figure 15.1[1]). Nevertheless, basic information can be conveyed easily and clearly in this form, even for a generalist analyst.

OVERLAYS

Overlays are used to display additional information about the terrain or activities within the geographic area bound by the map. Overlays can be used for both maps and aerial photographs. An illustration of a map overlay, from the U.S. Army's field manual entitled *Map Reading and Land Navigation*,[2] is shown in figure 15.2.

Figure 15.1.   First World War military intelligence students studying aerial photographs showing the intricate trench systems used by opposing forces.
Courtesy of the US Army

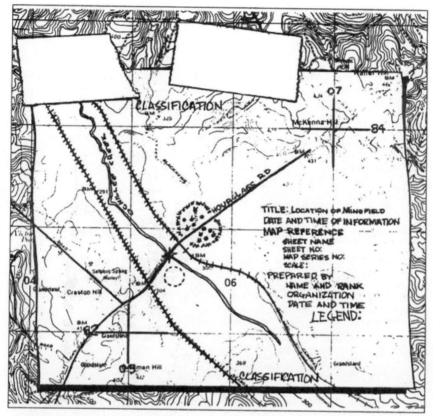

Figure 15.2.   Example of a military map overlay with notation in the margins.
Courtesy of the US Army

Physical overlays are usually clear sheets of plastic-like material that allow the analyst to hand draw or annotate the map/photograph with standardized symbols or codes. They are most commonly used in operational theaters that are characterized by quick decisions and frequent changes of plans. Marks on the clear sheets can be erased and redrawn as events or plans change. Several overlays can be used to "build" a picture of various activities or present different tactical options.

Anything that can be presented in a hand-drawn overlay can be presented electronically using a software package. This method is most effective in oral briefings where an image is shown, and with the click of a mouse button, additional information can be overlaid, thus emphasizing the point the analyst wants to make.

MOSAICS

A mosaic is created by combining two or more overlapping graphics in such a way that they form a single picture. Graphics can be maps, aerial photographs, and vertical photographs (including those taken at low and high oblique). However, the majority of mosaics are of a photographic nature, as they offer operational commanders and field operatives a panorama-like view of an area under investigation.

Since their employ during World War I, mosaics have been useful in displaying spatial data and, hence, vital intelligence. With today's software packages, photographs can be combined with terrain data so that a digitized image can be constructed to produce a three-dimensional, mosaic-orthogonal map.

These mosaics are valuable in training field operatives in such techniques as border crossing and covert insertions—techniques that require the operative to enter a hostile country (or in the case of law enforcement, a violent or dangerous neighborhood in an urban area). These computer-generated, mosaic-orthogonal maps can be extensive, covering tens of thousands of square kilometers.

KEY WORDS AND PHRASES

The key words and phrases associated with this chapter are listed below. Demonstrate your understanding of each by writing a short definition or explanation in one or two sentences.

- Aerial photograph;
- Map;
- Mosaic;
- Mud map; and
- Overlay.

STUDY QUESTIONS

1. Define the term *map*, and give three examples of commonly used maps.
2. Describe a situation where an intelligence analyst might use an overlay.
3. List the main features of a mosaic.

LEARNING ACTIVITY

Suppose you have been asked to construct a map overlay to assist field operatives in a covert operation. Using a street map for the city in which you live as an example, use some imagination in constructing an overlay that shows the following detail: a muster point for preoperation briefings, the location of a notional target, any potential hazards (countersurveillance points), and an exfiltration route for the covert team. Annotate the overlay with an appropriate security classification.

NOTES

1. John Patrick Finnegan, U.S. Army Intelligence and Security Command, U.S. Department of the Army, *Military Intelligence: A Pictorial History* (Arlington, VA: Department of the Army, 1984), 40.

2. U.S. Department of the Army, *FM3-25.26: Map Reading and Land Navigation* (Washington, DC: Department of the Army, 2001), figure 7-2.

16

🌀

Quantitative Analytics

This topic presents the essential statistical techniques used in intelligence analysis:

1. Levels of data measurement;
2. Univariate analysis;
3. Calculating percent increases and decreases;
4. Per capita calculations;
5. Index numbers;
6. Ratios;
7. Rounding numbers;
8. Bivariate analysis;
9. Statistical significance; and
10. Problem solving using algebraic equations.

LEVELS OF DATA MEASUREMENT

In intelligence research, if variables are observed, the data gathered by these observations are organized according to numbers attributed to them by analysts. Assigning numbers to represent variables is done as a way of preparing these data for statistical testing. The numbers assigned to the data are the factors that determine the *level of measurement* and, hence, the kinds of statistical tests that can be applied.

There are four levels of measurement: nominal data, ordinal data, interval data, and ratio data. Each level of measurement represents an increase in the types of statistical tests that are permissible. As such, the level of measurement, it could be argued, is the foundational assumption for all statistical testing. That is, the data type determines which statistical test can be conducted. If this assumption is violated, it makes the results of any subsequent statistical test invalid.

For instance, if data were of a nominal scale, then only statistical tests designed to analyze these types of data can be used (say, for example, chi-square analysis). However, if the data were of a ratio scale, then any test at the ratio level and those for lower-level data could be applied. This is because ratio data can be reduced to a lower level of measurement to be analyzed using an appropriate test. If the data are already at a lesser level of measurement, they cannot be converted to a higher level. Likewise, once the data are converted to a lower scale, unless the original data are retained in ratio form, they cannot be disaggregated.

Analysts also use what is termed *derived data*—but this is not the same as a level of measurement; it is information that is produced from combining or interrelating two or more data items to produce a new piece of information. For example, suppose an analyst has data on several events and also has data on when those events took place. Using a computer program, the analyst can produce new data based on an algorithm that, for example, requests "a list of all events that occurred before a certain date but after another specified date, as long as the event was not followed at any time by another type of event."

Nominal Data

Nominal data is the lowest level of measurement and comprises observations that can be placed in a group, for instance: Americans, Australians, Britons, Canadians, or New Zealanders (as such, it is sometimes referred to as a *categorical* scale). The chief attributes of nominal data are that there is no rank or order to the data—that is, one group is not "greater" or "more" than another group.

By itself, there is no "distance" between the groups as in higher-level data. So one cannot say that if you are an Australian, you are twice as "ethnic" as a New Zealander. All that can be said in this example is that one is in the Australian group, and the other is in the New Zealand group. Also, an observation can only be in one group, not multiple groups.

Ordinal Data

Ordinal data has the same attributes as nominal data, but in addition, it introduces the attribute of rank (sometimes referred to as *rank scale*). Rank in this sense suggests that the scale has some direction—say, from less to severe. Take, for instance, the crime of terrorism—three events could be ranked as to their severity: kidnapping, assassination, and bombing. One will note that they have the same attributes as nominal scale data in that observations can be placed in groups, but these groups have a relationship in how they are ranked. Because these observations can be ordered in this way, the level of measurement increases.

Nevertheless, even though there is direction to these data, there is no indication of distance between the data items. For instance, an analyst cannot say that assassination is twice as severe as kidnapping. However, the analysts can say that bombing is greater than kidnapping, that bombing is greater than assassination, and that assassination is greater than kidnapping.

Interval Data

Interval data has the same attributes as ordinal data, but it introduces the dimension of distance, or *interval*. That is, data measured using this scale will be able to demonstrate a common unit of measurement for all observations.

Take, for instance, the interval measure of time—a terrorist planted an improvised explosive device in a busy marketplace at 1:00 p.m., and it was detonated by remote control at 1:30 p.m. The analyst can conclude that there were two distinct events (nominal scale), one event was more serious than the other (on the ordinal scale, it could be argued that planting a bomb is serious, but detonating it is far more serious), and the time that elapsed between the two events was thirty minutes (interval scale). Because there is now a measurable distance, the interval scale allows the analyst to conduct arithmetic calculation on the data. One should note that there may be a zero point in this scale (e.g., 00:00 hours) but that zero is an arbitrary notion—it is not real.

Ratio Data

Ratio data is the highest level of data, as it has all the attributes of interval data, but in addition, it has a real zero for the scale's starting point. Having a real zero starting point reference allows the analyst to calculate ratios between any two observations—data can be added, subtracted, multiplied, and divided.

By way of example, suppose the terrorist-improvised explosive device cited above was estimated to contain fifteen kilograms of high explosive. Suppose also that another explosion that day was due to a device containing thirty kilograms. The analyst can conclude that there were two explosions (nominal scale), one was bigger than the other (ordinal scale), the larger explosion was fifteen kilograms greater (interval scale), as well as having twice the destructive power (ratio scale, assuming in this example that destructive power is linear).

UNIVARIATE ANALYSIS

Univariate analyses are used when the analyst wants to simply describe a person, organization, location, or object that consists of a single dependent variable. Hence, univariate analysis is also known as *descriptive statistics*. Contrast this type of analysis with bivariate and multivariate analyses where two (bi-) or more (multi-) variables are analyzed (in such analyses, the variables may or may not be dependent upon each other). Univariate analysis can be used with the different levels of data as shown in table 16.1. The descriptive statistics that can be produced are listed in the right-hand column.

Frequencies

Constructing a frequency distribution is often an analyst's first task in analyzing data. This is done by counting the number of observations per category (nominal and ordinal level) or per score (interval and ratio level). The results can be described in the intelligence report's narrative as well as being displayed graphically. Figure 16.1 (ordinal level) and table 16.2 (ratio level) are examples of how terrorist event data can be presented. Figure 16.1 shows the number of terrorist events by organization before and after the 1986 U.S. air raid on Libya,[1] whereas table 16.2 shows the number of (notional) bombing events (X) terrorist groups were responsible for in Country Q.

Count

Count is the total number of values in a distribution (i.e., a data set). Using the following distribution as an example, the count would be

Table 16.1.   Examples of Descriptive Statistics

Measurement Level	Analytic Technique
Nominal	Frequencies
Ordinal	Frequencies, Range, Median, Mode
Interval	Frequencies, Range, Minimum, Maximum, Median, Mode, Mean, Weighted Mean
Ratio	Frequencies, Range, Minimum, Maximum, Median, Mode, Mean, Weighted Mean

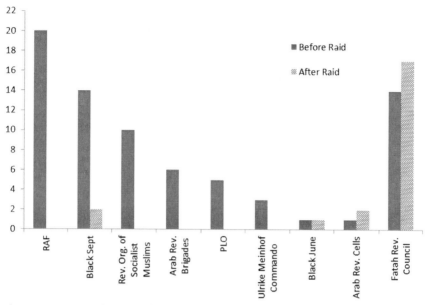

Figure 16.1. Terrorist events by Libyan-sponsored groups targeting Americans and U.S. property abroad.

eleven. These data could represent the number of people killed by roadside bombs, or any number of other intelligence-related events:

4, 8, 10, 12, 15, 16, 19, 20, 24, 28, 31

Minimum and Maximum

The minimum is the lowest value in a distribution. The maximum is the largest value. For instance, using the same data set directly above, the

Table 16.2. Frequency Distribution of a Series of Notional Bombing Events

X (score)	f (frequency)	fX
5	2	10
4	3	12
3	7	21
2	15	30
1	30	30
	N = 57	ΣfX = 103

minimum number of people killed by roadside bombs, the minimum and maximum are 4 and 31, respectively:

$$4, 8, 10, 12, 15, 16, 19, 20, 24, 28, 31$$

Range

The range is the difference between the minimum and maximum values within a distribution. This is calculated by subtracting the smaller value from the larger. Using the same example above, the range would be 27 as calculated here:

$$31 - 4 = 27$$

Mean

The *mean*, also known as the arithmetic *average*, is used to average out quantities. It is calculated by adding all the values of a data set, then dividing the sum by the count of those numbers. Again, take for instance the distribution above: the total number of people killed was 187. Divide 187 by 11, and this equals 17:

$$(4 + 8 + 10 + 12 + 15 + 16 + 19 + 20 + 24 + 28 + 31) = 187 / 11 = 17$$

The mean can be used with both interval and ratio data; however, one of the disadvantages of using the mean is that it is affected by extremes at either end of the distribution. For example, if we use the notional road-side bomb data set and substitute 97 for the value 31, the resulting mean would be skewed to 23. Nevertheless, the mean is a useful statistical test, as it can be used to conduct further tests—for instance, an analyst can compare the means of several different samples.

Weighted Mean

Another method used to average quantities is the *weighted mean*. It is used in cases where not all of the quantities are of equal importance. The formula for calculating a weighted mean is expressed in the following, where "bar-x" is the mean, w is the weight, x is the number, and Σ is the sum:

$$\bar{x}_w = \frac{w_1 x_1 + w_2 x_2 + w_3 x_3 + \ldots + w_n x_n}{w_1 + w_2 + w_3 + \ldots + w_n} = \frac{\Sigma\, w * x}{\Sigma\, w}$$

Suppose a situation where an intelligence analyst is trying to establish the average price paid for heroin on the street but on a national basis.

Data from each state's capital city are compiled and are weighted according to the population of each city (i.e., using census data). This takes into account the fact that there are price variations at the same point in time between locations because illicit drug prices increase "as one moves away from the drug sources and prices are lower in larger markets."[2]

Suppose that there are three capital cities in Country Q. The price paid for heroin in one city is $148 per gram, another is $256, and the third is $300. The populations of the cities are 5 million people, 2 million people, and 1 million people, respectively. Therefore:

$$\bar{x}_w = \frac{(5)(148) + (2)(256) + (1)(300)}{5 + 2 + 1}$$

$$= \frac{1{,}552}{8}$$

$$= \$194 \text{ per gram}$$

If the analysts had averaged these data without using a weighting system to take into account the price variations caused by distances from the point of importation (the largest city with its port and international airport), they would have obtained $235. This is because the state with the smallest population that was far removed from the large port city—and hence had the highest price—skewed the average upward.

Weighted Grand Mean

There may be times when a *grand mean*, or the mean of the sum of all the means, is needed to be calculated. Such cases may arise where, for instance, a mean for heroin in each state's capital city has been calculated, but the national average is needed. To do this, the analyst totals the means for each city and then divides this figure by the number of cities. But because averaging the average can skew the results (sometimes considerably), a *weighted grand mean* is used. This is calculated by using the same formula for a weighted average; however, instead of using *x*, the analyst uses *bar-x*, that is, the mean.

Median

The median is the middle value of a distribution. It is, therefore, the halfway point between those values that are greater than the median and the half that are less than the median. Using the following data set, the median is 16:

4, 8, 10, 12, 15, 16, 19, 20, 24, 28, 31

The median is less susceptible to extremes at the edges of the distribution. Again, take for example the notional roadside bomb data set we used previously, and again substitute 97 for the value 31, as was done for the mean, but now the resulting median would be 16, not the skewed 23 that was seen under the mean. Therefore, the median is generally more useful where there are extremes, and the mean most useful when the distribution is absent of such features.

Mode

The mode is the most frequently occurring value in a distribution. However, some data sets do not have a mode, as is evident with the example distribution used so far—there is a single count for each number. If, by contrast, the following distribution was used, the mode would be 12:

$$4, 8, 10, 12, 12, 15, 16, 19, 20, 24, 28, 31$$

Sometimes data sets have several sets of numbers appearing in equal frequency. In such cases, these are referred to as being bimodal (two modes), trimodal (three modes), or in the case of more—multimodal.

Bimodal:

$$4, 8, 10, 12, 12, 15, 16, 19, 20, 24, 28, 28, 31$$

Trimodal:

$$4, 8, 10, 12, 12, 15, 16, 19, 19, 20, 24, 28, 28, 31$$

Of the three measures of central tendency—mean, median, and mode—the mode is the least useful unless the number of values represented by the mode form a large percentage of total distribution. Moreover, no further analysis can be conducted on the mode—an analyst cannot, for instance, compare modes of different samples, as the result would make no sense at all.

CALCULATING PERCENT INCREASES AND DECREASES

Sometimes an analyst will need to express a number as a fraction of 100. This is usually done because the analyst needs to show how one quantity is related to another. The mathematical function used to do this is *percent* (meaning, per hundred). As an illustration, 54 percent is 54 / 100, or expressed as a decimal, 0.54. When a number increases, an analyst can calculate its percentage increase using the following formula:

percent increase = [(*new figure* − *original figure*) / *original figure*] × 100

When a quantity decreases, the analyst can calculate the percentage decrease by:

percent decrease = [(*original figure* − *new figure*) / *original figure*] × 100

It is important to note that both the formulas exhibit the following principle:

percent increase / decrease = (*change in figure / original figure*) × 100

This is because the analyst is calculating either the percent increase or decrease with the amount of change to the original figure as demonstrated in the following two examples:

1. The number of people attending political rallies in the city of Orrenabad rose from 16,000 to 22,000. What percent increase does this represent?

   percent increase = (22,000 − 16,000) /
   16,000 = 6,000 / 16,000 = 0.375 × 100 = 37.5%

2. The number of political websites hostile to the new democratically elected government in Orren declined from 930 to 200. What percent decrease does this represent?

   percent decrease = (200 − 930) /
   930 = −730 / 930 = 0.785 × 100 = −78.5%

PER CAPITA CALCULATIONS

Per capita simply means per person. An analyst can calculate the per capita occurrence of x where x is the issue under investigation:

per capita = (*total x / population*) x *rate*

So, using an example of suicides, the per capita figure would be the total number of self-inflicted deaths divided by the population. In this case, the population could be either the population of the town or city, the region or state, or the country. This statistic can then be used as a means of comparison to other towns or cities or from country to country in a meaningful way.

One of the detractions of per capita calculations is where the figure is very small; in such cases, the ratio becomes meaningless. In such cases, the analyst can use per 100,000 in the population, 10,000, or even 1,000 as alternatives.

By way of example, take the notional regional town of Isobel. The town recorded 200 residential burglaries during the year, and it had 100,000 households. Therefore, the burglary rate would be 0.002. This is not a very meaningful figure. However, if it is calculated per 1,000 in the population, then it is 2 burglaries per 1,000 households.

$$(200 \ / \ 100,000) \times 1,000 = 2$$

By using a rate per 1,000 instead of a per capita rate, in this example, the figure can more easily be understood.

INDEX NUMBERS

A benchmark can assist analysts when comparing changes in different kinds of data, and in statistical terms this is referred to as an *index*. "Since index numbers show percent changes rather than arithmetic change, the size of the data and the units of measurement are not important."[3]

As a demonstration, the rate of inflation can be compared with the production of goods by using the consumer price index with the industrial production index. "Although prices are measured in dollars and production is measured in physical volume (number of cars, tons of coal, etc.),"[4] the two can be easily compared using this method.

As part of a study into heroin trafficking in Australia, Prunckun[5] used index numbers to control for confounding variables. His study collected data relating to Australia's population, its rate of inflation, and its number of sworn police officers. These data were gathered as a means of comparing or adjusting the rate of change observed in variables. For instance, population data were used to adjust methadone and heroin usage rates relative to "the population in the age groups in which opiate users are found (namely 15 to 44 years)."[6] The inflation rate was helpful in adjusting the illicit drug prices, and the police staffing data were helpful in comparing changes in police numbers with that of some of the law enforcement workload data. The formula for calculating an index is:

index = (current value / base value) $\times$ 100

Suppose that Country Q had 5 warships in 2010. By 2015, intelligence indicated that this number had grown to 20 (with the intervening years being 6 in 2011, 9 in 2012, 14 in 2013, and 17 in 2014). How has this number changed between 2010 and 2015?

The base period is 2010, and the *base value* is 5, which is given a value of 100 that corresponds to 100%. The index is then calculated for the period following the base period using the above formula and is displayed in table 16.3. The index number shows that there was an increase of 300% in warships since the base period (i.e., 400% − 100% = 300%).

RATIOS

An analyst who wants to compare two things can use a ratio. A ratio is an expression of something *to* something else: for instance, a ratio of small arms to soldiers. Ratios are expressed in several ways: either as a fraction (3/4) or using the word *to* (3 to 4) or using a colon (3:4).

To illustrate this, say an analyst observes that there are 90 soldiers in a combat force of a foreign country that are engaging friendly forces involved in counterinsurgency operations. Three are assigned to radio communications. Later, signals intelligence indicates that this foreign force has added a fourth radio operator, but the size has grown to 120 soldiers. Despite the changes in overall numbers, the ratio of radio operators to soldiers within the force has remained the same—30:1 before, and 120:3 after (which can be simplified by mathematical reduction to 30:1).

ROUNDING NUMBERS

Rounding numbers is commonly used where the decimal fraction makes little sense or adds confusion. Take for example a hypothetical case where an analyst calculates the average number of rounds fired per soldier during the last six skirmishes with enemy forces was 194.6342. At four decimal places the result adds no discerning insight to the study's findings, as most people cannot think to four decimal places (we usually can visualize one or two decimal places, but no more—e.g., 0.5 is half, and 0.25 is a quarter, and 0.75 is three-quarters). Besides, analysts should be asking

Table 16.3.   **Example of Index Numbers for Warships**

Year	Number of Warships	Index Number
2010	5	100
2011	6	120
2012	9	180
2013	14	280
2014	17	340
2015	20	400

themselves what insight this level of detail will add to the conclusions. Sometimes an estimate is all that is needed to demonstrate a point being made or make a prediction. In these cases, rounding is advisable. Step by step:

1. Determine the unit that is required to be rounded off—thousands, hundreds, units, or a decimal fraction;
2. Examine the number to the right of the unit to be rounded off;
3. If this number is 5 or more, add an additional unit;
4. If the number is less than 5, subtract a unit.

For instance, using this mathematical convention, if an analyst was rounding 194.6 to the nearest unit, it would be rounded up to 195. If rounding 194.6 to the nearest ten, it would be rounded down to 190. But if the analyst was rounding 194.6 to the nearest hundred, it would be rounded up to 200.

BIVARIATE ANALYSIS

Chi-square

Chi-square is one of the more useful statistical tests for intelligence analysts because it can be used on nominal-level data. It has few assumptions and is, therefore, straightforward to apply and interpret.

Chi-square is a nonparametric test of statistical significance. The term *nonparametric* refers to statistics that deal with variables that are without assumptions as to their form or their distribution parameters. It returns a statistic that reflects the "goodness of fit" or the difference between the observed frequency and the expected observations according to a model hypothesis (H_0—known as the *null hypothesis*).

The power of the chi-square statistic is that it will tell the analyst whether the actual distribution occurred by chance or was likely to be the result of the interaction of the independent variable according to a level of confidence (e.g., .01 or .05). Later in this chapter, the importance of setting the appropriate level of confidence is discussed in more detail. There are only three requirements for using chi-square, and these are:

1. Any level data can be used;
2. There must be more than five observations per category (if there are less than five observations, they will be meaningless); and
3. The observations must be independent.

Unlike some other statistical tests, there is no direction indicated by chi-square, so once the result is obtained, an inspection of the data is required to determine the direction (that is, whether it is positively correlated or negatively correlated). Consider the following example of a single sample chi-square: Analysts are concerned about whether the numbers of terrorists in a target organization's cells in North America are equal to those in the organization's Asia region. The observed frequencies are arranged in table form (referred to as a *contingency table*). If the data were represented by two independent samples (that is, in the form of a 2 × 2 contingency table), then the convention is to list the independent variable (x) along the top of the table with the dependent variable along the side (y).

Chi-square can also be used for cases involving two independent samples. For instance, suppose an analyst wanted to know if there were significant differences between insurgents who were young males as opposed to older males and whether these people had previously been involved in criminal activity before joining the insurgency. The null hypothesis (H_0) would be: The number of terrorists in North American cells and Asian cells are equivalent. The alternative hypothesis (H_1) would be: The number of terrorists in Asian cells is more. The observed distribution is:

North America	86
Asia	120
Total	206

The expected distribution under the H_0 is for 50 percent to appear in each:

North America	103
Asia	103
Total	206

The formula for calculation of chi-square is as follows:

$$X^2 = \Sigma \frac{([A_1 - E_1] - .5)^2}{E}$$

X^2 = chi-square
A = actual or observed frequency
E = expected frequency
$-.5$ = Yates' correction for continuity

The calculations follow; however, many software spreadsheet packages contain the chi-square function. It is a simple matter of entering the data into the spreadsheet, indicating the degrees of freedom, and activating the chi-square function from the menu options. The spreadsheet will return the chi-square statistic (here it was done manually and is shown in table 16.4).

The degrees of freedom are calculated thus: $df = C - 1$ (where C represents categories). So, in this example it would be $2 - 1 = 1df$.

Table 16.4.   Manual Calculations for Determining Chi-Square

	North America	Asia	Total
Actual/Observed	86	120	206
Expected	103	103	206
$(A-E)-.5$	16.5	16.5	
$([A-E]-.5)^2$	272.25	272.25	
$([A-E]-.5)^2/E$	3.17	2.27	
Σ	5.44		

Yates' correction for continuity is applied in instances where the degrees of freedom are equal to 1 or where the observed frequencies are less than 10. This prevents an overestimation of statistical significance.

In chi-square analysis, the *critical value* is the threshold that determines at what point an analyst would not reject the null hypothesis. Using the critical values contained in the appendix, we observe that a value of 5.28 is greater than the required 3.84 at the .05 level (1*df*), so we can reject the null hypothesis—numbers of terrorists that exist in the Asia cell are greater (i.e., we accept H_1—the alternative hypothesis). This means that there would be less than 5 chances in 100 that a result like this would be obtained if random variation was the only explanation.

Note that the result in this example does not *prove* the alternative hypothesis, it merely *supports* the explanation. However, a hypothesis can be disproven. It is important that these subtleties are understood and reflected accurately in any written report or oral briefing based on such findings. (See also the discussion under "Statistical Significance" below.)

As can be deduced from the formula, the chi-square statistic is the product of each of the individual frequencies. Therefore, each frequency contributes to the final chi-square statistic. If a given frequency is greatly different from the expected frequency, then its contribution to the chi-square statistic can be expected to be large. Conversely, if the frequency closely aligns itself to the expected frequency, then the contribution of that frequency to chi-square will be small.

It is clear that large chi-square statistics indicate that the contingency table contains a frequency(s) that differs noticeably from the expected frequency(s), but the statistic will not be able to point to the frequency(s) responsible for the elevated chi-square. It can only indicate that such a frequency(s) is present. When an excessive result is obtained, it requires the analyst to examine the table in order to determine which frequency(s) is the cause.

The body of the analyst's report should discuss the conclusions drawn from these results, what they mean, and the implications they have for the study. When discussing these issues, refer directly to the table(s) so the reader is not left in doubt about any aspect of the conclusions.

Scatter Plots

A *scatter plot* (sometimes termed *scattergram* or *scatter diagram*) is used to visualize relationships between two data sets (i.e., paired samples). In doing so, the two variables do not have to have a dependent and independent relationship, as plotting will demonstrate the degree the variables are correlated. As such, plotting data using this method is synonymous with *regression analysis*.

Scatter plots are able to reveal a number of relationships in addition to the degree of correlation: for instance, a positive relationship, a negative relationship, and a null relationship. These are illustrated in the following series of figures.

Figure 16.2 shows the pattern of dots sloping from the lower left to the upper right, thus demonstrating a positive relationship. Figure 16.3 shows a pattern of dots sloping from the upper left to lower right, suggesting a negative correlation. The "tighter" the dots are gathered in forming a straight line, the more correlated the data are—see figure 16.4 for an example of a perfectly corrected paired sample. Finally, figure 16.5 shows a nonlinear, or null, relationship.

An analyst can quickly construct a scatter plot either by hand or by using a software package (usually a spreadsheet program). The first step

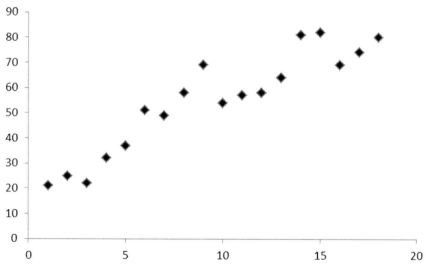

Figure 16.2. Positive correlation

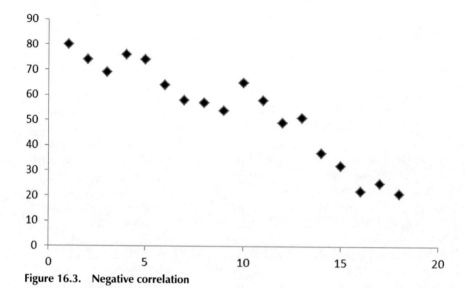
Figure 16.3.   Negative correlation

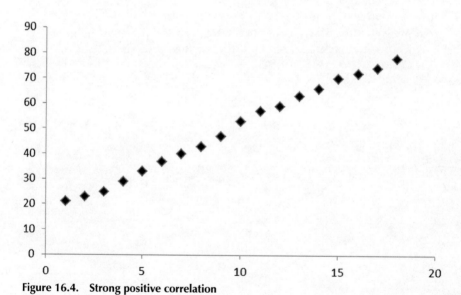
Figure 16.4.   Strong positive correlation

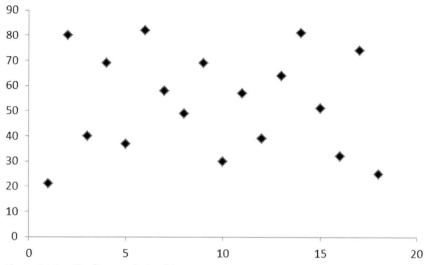

Figure 16.5. Nonlinear relationship

is to draw a chart with a horizontal axis (x) and a vertical axis (y) and place appropriate and proportional numeric gradations along the axes. Data are then plotted on the graph. If doing this by hand where values are repeated, the analyst can indicate the repeated values by placing a circle around those points. If using a software package, the program will generate the chart once the data items are entered into the spreadsheet. Follow the menu options to produce the scatter plot.

The chief advantage of a scatter plot is that it will indicate the strength of the relationship between two variables. The stronger the relationship the more likely a change in one variable will result in a change in the other.

Note that these results are not able to predict cause. All they are able to say is that there is a relationship (or not) between the two sets of data. However, by examining other factors associated with the issue under investigation, it may be possible to make such a judgment.

Take for instance the simple example of where it is observed that when armored vehicles are required to carry additional ammunition, the amount of fuel these vehicles consume increases. This is a concern, as it shortens the distance they can patrol. A scatter plot confirms a strong positive relationship between the two variables—the more ammunition loaded onto a vehicle, the more fuel it consumes (and this might look like the scatter plot shown in figure 16.4). But in order to draw the conclusion that this additional ammunition has *caused* this, other factors need to be assessed, and all these factors need to be framed in a reasoned argument: ammunition has mass, and this mass increases the weight of the vehicle. Other studies have shown that additional weight increases a vehicle's fuel consumption, and less fuel means fewer kilometers can be traveled. It therefore can be concluded that the additional ammunition is the cause of the reduced distances that armored vehicles can patrol.[7]

STATISTICAL SIGNIFICANCE

Type I and Type II Errors

The concept of *statistical significance* is important to the analyst, as it makes clear what likelihood there is for an error in judgment. At the center of this process is the *confidence level* chosen by the analyst—this is referred to as the *alpha-level* or *p-value*, for example: .05, .01 (or another level either higher or lower). Regardless of the level chosen, there is the risk that the null hypothesis could be rejected when it should have been accepted or vice versa—accepted when it should have been rejected. The risk of error increases as the confidence level gets smaller.

For instance, a *p*-value of .01 means that there would be less than one chance in one hundred that a result like this would be obtained if random variation was the only explanation (i.e., a 99% chance of being true). A *p*-value of .05 indicates that there is a 95% chance of being true. And, a *p*-value of .001 indicates there is a 99.9% chance of being true.

Such errors are referred to as type I errors and type II errors. The former are *false positives*, and the latter are *false negatives*.

As pointed out in the section above about chi-square, these results do not *prove* the alternative hypothesis; they merely *support* it as an explanation. That is to say, with a confidence level of, say, .05, there is still a 5% chance that the findings may have been the result of another factor(s). However, the hypothesis can be disproven if it fails to meet the confidence level.

By way of example, suppose an analyst has selected a confidence level of .05; in this case, the analyst could more easily reject the null hypothesis and declare that there is a statistically significant difference in the data than what would have occurred if he or she had selected the .01 level. However, the analyst would be wrong in this conclusion 5 percent of the time. If, however, the analyst selected a confidence level of .01, he or she would be wrong only 1 percent of the time.

Type I errors are when some phenomenon is observed and the conclusion is drawn that there is a difference when in fact none exists—this is why it is viewed as a false positive. A type II error is a false negative—accepting the null hypothesis when it should have been rejected. This is where the analyst fails to observe a difference and accepts the null hypothesis. The relationships between these decisions are summarized in table 16.5.

What is the impact on the analyst's findings of type I and type II errors? A type I error could be described as a "false alarm"—sending a signal that the observed phenomenon is worthy of note (and presumed action) when in fact it is a mistaken conclusion. The analogy could be a miner finding pyrite (i.e., fool's gold) instead of gold. In contrast, a type II error could be attributed to a lack of sensitivity of the data collection instrument or an omission of some sort so that the difference was not detected. If the miner analogy is used again, the miner would have thrown out the gold with the tailings from the mine.

From this discussion, the analyst can see that in any research project, they are always at risk of making one of these two types of errors. So, which error is "worse" to make? It depends on the question that has been asked by the decision maker. Both can be equally devastating as can be seen in the following two examples:

- If the decision maker asked if there are any weapons of mass destruction in Iraq, then committing a type I error will have serious repercussions if the decision maker is planning a military invasion (i.e., wrongly concluding that such weapons existed when there were none).
- If, however, the analyst is asked whether the United States faces an attack by an international terrorist group (e.g., in the manner of al-Qaeda), then a type II error needs to be avoided—overlooking a genuine threat to national security could result in another September 11, 2001–style attack.

PROBLEM SOLVING USING ALGEBRAIC EQUATIONS

Algebraic equations are used in a range of research activities to solve a myriad of problems. In essence, algebraic equations are formulae-based

Table 16.5. Summary of Statistical Decisions

	Where H_0 is true	Where H_0 is false
If H_0 is rejected	Analyst commits a type I error (false positive)	Analyst's decision is correct
If H_0 is not rejected	Analyst's decision is correct	Analyst commits a type II error (false negative)

models that represent situations in the physical world. Rather than try to canvass the countless ways algebra can be used, we will draw on an indicative example to demonstrate the power of this type of analysis.

Consider a situation where an operational commander needs to be advised as to the estimate of the arrival time of enemy troops (in a military intelligence context) or the arrival of a motorcycle gang that is on the move (in a law enforcement intelligence context). Using algebra to solve this dilemma becomes an exercise in predictive intelligence, albeit one that is purely operational (i.e., a road movement calculation). To solve this problem the analyst:

- Notes the location of the troops/motorcycle gang at present;
- Notes the location of where the troops/motorcycle gang are expected to arrive;
- Calculates the distance between the two points;
- Notes the rate of advance (e.g., from information received from a reconnaissance unit, an airborne observer, or covert surveillance);
- Calculates the approximate arrival time for the troops/motorcycle gang, taking the distance between the two location points and dividing it by the rate of advance ($T = D / R$).

In this example, suppose the distance between the two points of interest is 200 kilometers. And suppose that the persons of interest are advancing at a rate of 50 kilometers per hour. Therefore, 200 km / 50 km per hour = 4 hours. The intelligence analyst can then brief the commander that the estimated arrival time of the persons of interest will be in approximately 4 hours.[8]

This type of estimate can be applied to business intelligence and private sector intelligence too. In the case of the former, analysts could estimate the market arrival of a competitor's new model Gizmo. If the rate of production per unit is known (or can be estimated based on previous information or similar processes), then the time frame for delivery can be calculated by inverting the equation used above and amending the variable labels to reflect the events:

delivery date = number of units needed × rate of production

In the case of private sector intelligence, an analyst can use algebraic equations to, say, deduce that it was feasible for the suspect in an insurance fraud matter to set the fire in question. Using the formula cited in the distance traveled example above, the analyst could use it to show that the person of interest was capable of traveling, without exceeding the speed limit, from his location to the site of the alleged arson and back within the times given by, say, witnesses.

KEY WORDS AND PHRASES

The key words and phrases associated with this chapter are listed below. Demonstrate your understanding of each by writing a short definition or explanation in one or two sentences.

- Algebraic equations;
- Average;
- Bivariate;
- Categorical data;
- Chi-square analysis;
- Descriptive statistics;
- Frequency;
- Index numbers;
- Interval data;
- Level of confidence;
- Maximum;
- Mean;
- Medium;
- Minimum;
- Mode;
- Multivariate;
- Nominal data;
- *P*-value;
- Per capita;
- Percent increase and decrease;
- Range;
- Ratio data;
- Ratios;
- Regression analysis;
- Rounding;
- Scatter plot;
- Statistical significance;
- Type I error;
- Type II error;
- Weighted grand mean; and
- Weighted mean.

STUDY QUESTIONS

1. What are the four levels of measurement in statistics? Describe the attributes of each level.

2. List the different types of univariate analysis available to the intelligence analyst.
3. Describe the differences between mean and weighted mean. Give an example of how an analyst might use each.
4. What is the difference between a type I error and a type II error? Explain.

LEARNING ACTIVITIES

1. Suppose ocean-going freighters regularly use the straits off the coast of Country Q to save travel time. Suppose also that these straits have recently become the subject of pirate attacks. You, as an intelligence analyst, have been asked to calculate the decrease in international shipping using these straits after the appearance of pirates. The daily number of ships before the pirate attacks was 416 and after was 232. What was the percentage decrease due to this hostile maritime activity?
2. Community elders in Country Q are concerned about attacks on street markets—they suspect that they are not random and that this could signify growing factional tensions. The intelligence cell attached to the nation's paramilitary police has divided the country into five equal regions and collated the attack data accordingly. This information revealed the following distribution:

Table 16.6.

Region	Number of Attacks
1	26
2	19
3	31
4	17
5	11

Using a chi-square test of significance, determine whether it is likely that this distribution of attacks is by chance (random) or that other forces may be at play. Use a .05 confidence level. What can you conclude from these results?

NOTES

1. Henry Prunckun, "Operation El Dorado Canyon: A Military Solution to the Law Enforcement Problem of Terrorism—A Quantitative Analysis" (master's thesis, University of South Australia, 1994), 47.

2. Jonathan Caulkins, "What Price Data Tell Us about Drug Markets," *Journal of Drug Issues* 28, no. 3 (Summer 1998): 602.

3. Ester H. Highland and Roberta S. Rosenbaum, *Business Mathematics*, third edition (Englewood Cliffs, NJ: Prentice Hall, 1985), 420.

4. Highland and Rosenbaum, *Business Mathematics*, 420.

5. Henry Prunckun, *Chasing the Dragon: A Quantitative Analysis of Australia's Law Enforcement Approach to Combating Heroin Trafficking—1988 to 1996* (PhD diss., University of South Australia, 2000).

6. Commonwealth Department of Human Services and Health, *Review of Methadone Treatment in Australia* (Canberra, Australia: Australian Government Publishing Service, 1995), 29.

7. Although this is a simple example, it assumes that no other factor has changed, just the additional ammunition. For instance, it assumes that the crew remain the same in number, that no other equipment is loaded with the ammunition—such as additional weapons to fire the ammunition—and so on. These types of factors are assessed by the analysts when making a judgment about cause and effect.

8. U.S. Department of the Army, *FM 34-2-1: Tactics, Techniques and Procedures for Reconnaissance and Surveillance, and Intelligence Support to Counterreconnaissance* (Washington, DC: Department of the Army, June 1991), 2–20.

17

§

Displaying Information in Figures and Tables

The key concepts that will be discussed in the chapter are:

1. Figures;
2. Tables; and
3. Stem-and-leaf plots.

FIGURES

Graphs, charts, photographs, drawings, diagrams, as well as other terms are all considered *figures* when writing an intelligence report. The term *figure* is a distillate of all graphical representations into one summary term. The only other term that is used in presenting research results is the term *table*. *Figures* and *tables* are the only two terms that should be used in intelligence writing. However, having said that, we will look at the various lay terms as a way of explaining how results could be presented.

Graphs are used to display numeric data pictorially; in effect, they are symbolic representations. In practice, the term *graph* is used interchangeably with the term *chart*. Intelligence analysts use graphs to display information that would be too complex to do so in narrative form. Therefore, the graph offers the analysts a simple, concise format for conveying the gist of their results to the decision maker.

The most commonly used (and easily recognized by nonanalytic users) are the *pie chart*, *bar chart*, and *line chart*. A pie chart is a circular figure divided into segments that resemble a pie. Each "slice" of the pie represents a particular data item. The size of the overall pie is 100 percent, with each slice being a representation of a proportion of the total.

A bar chart uses vertical bars (but sometimes they are laid out horizontally) to show the relationship between data across categories. Bar charts are used where the data items are categorical (i.e., independent of each other): for instance, the number of rockets, automatic weapons, and hand grenades that a guerrilla force may have in its possession (a pie chart could be used for the same purpose, as it also represents individual data items). The bars are not drawn contiguously, as in a histogram, but with separations between them to demonstrate they are not related by time or score interval. If, however, the bars are joined in a contiguous manner, then it is referred to as a *histogram* (see figure 17.1). A histogram is a picture of a grouped distribution, which shows the shape of the distribution.

A *frequency polygon* serves the same purpose as a histogram, but it only shows the midpoint of the intervals of the scores. It can be displayed as either bars or dots. If shown as a series of dots, each dot is connected by a line (see figure 17.2). In this regard, the line becomes a bar chart but with the bar image suppressed—they are merely represented by dots joined by a line.

In contrast, a line chart is used where the data are required to be displayed showing passage of time (e.g., a time series study), say, the number of rocket-propelled grenades an insurgent group has held in its cache month by month (see figure. 17.3).

Some forms of graphs use pictures or symbols in place of bars, columns, or lines as a way of reinforcing the message. These are termed *pictographs* and lend themselves to briefings where the analyst is presenting his or her results via a data projector on a large screen. However, before

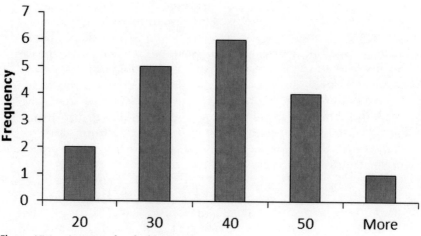

Figure 17.1.   An example of a histogram.

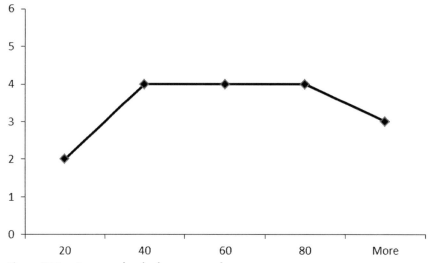
Figure 17.2. An example of a frequency polygon.

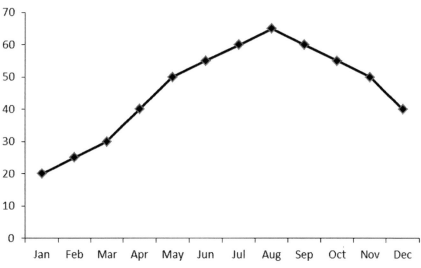
Figure 17.3. An example of a line chart.

an analyst uses such a technique, thought should be given to whether the inclusion of pictures will trivialize the presentation. Images that are provided via some commercial software packages could be construed by decision makers as frivolous, and therefore, their use in a presentation may lessen the impact of the message being conveyed.

SOME TIPS FOR CONSTRUCTING GRAPHS

- Give graphs captions that reflect the dependent and independent variables being displayed.
- The X axis displays the independent variable, while the Y axis the dependent variable.
- Label both axes and assign an appropriate unit of measurement.

X and Y Axes

Remembering which axis X refers to and which Y refers to is a common problem, even for seasoned analysts. There are many rhymes to help remember, and here is one that is often used: "X to the left; Y to the sky." That is, the X axis appears on the left of the figure and runs left to right (i.e., the horizontal axis), whereas the Y axis appears at the base of the figure and runs from the bottom to the top (i.e., the vertical axis).

Traps to Avoid

Here are a few things to keep in mind when constructing graphs:

- Do not be tempted to enhance a graph by using what could be considered a gimmicky feature contained in some software packages—such as three-dimensional charts. Features like this trivialize your research and distract the decision maker's attention away from your study's results and focus too much on the eye-catching graphical image;
- Be mindful of inadvertently distorting chart results by setting the baseline value (i.e., the value at the bottom of the vertical, or y, axis) to a score other than zero; and
- Do not mistakenly display categorical data as a frequency polygon (i.e., along the horizontal, or x, axis)—instead, present these data as a bar chart.

TABLES

A *table* is a data set arranged in columns. Like graphs, tables make the presentation of numeric data easy for the reader to understand. The information contained in a table needs to be simply laid out and concisely labeled so the reader is clear about the points being demonstrated.

When constructing a table, convention is to only use horizontal lines. Some software packages place vertical lines into tables, but strictly speaking, this is not correct. An example of a properly laid-out table is shown in table 17.1.[1] Each column has a descriptive header that identifies the data in each column. Each row also has a header, identifying the data in the row. The data appears in the remaining central part of the table. The caption appears either at the top or bottom of the table.

Because we read from left to right, it makes sense to lay the table out in that order, as the reader is likely to scan the table beginning at the left and moving to the right; and from the top, moving down to the bottom. Note in the example in table 17.1 that the "before" data appears to the left of the "after" data for this reason (i.e., in chronological order).

Column and row spacing is important for a pleasing visual presentation as well as for understanding. Columns and rows should be equally spaced with enough "white space" to make reading easy. Large numbers should have commas as separators for thousands, and decimal places kept to a minimum. Missing data in table cells should have some explanation, such as "not available."

Consider rounding very large numbers for ease of understanding for readers who find dealing with numbers difficult. For instance, if a table contains several columns of numbers, they could be rounded for clarity— e.g., 1,239,867 could be rounded to 1,240,000. You can see that rounding makes for ease of understanding and comparison of data within the table. At the same time, it in no way jeopardizes the overall integrity of the information being presented.

Table 17.1.   Worldwide Terrorist Events by Severity Level Pre- and Post-1986 U.S. Air Raid on Libya

	Terrorist Events		
Severity Levels	*Before Raid*	*After Raid*	*% Change*
High	541	458	−15.3%
Medium	170	155	−8.8%
Low	145	244	+68.3%
Totals	856	857	+00.1%

STEM-AND-LEAF PLOTS

As an alternative to displaying data in a histogram, the analyst can use a *stem-and-leaf plot*. Stem-and-leaf plots are constructed by creating a *stem* (left-hand column) that contains the tens-unit digits. The *leaves* of the plot are the integer-unit digits for each corresponding tens-unit and listed in the right-hand column. Once constructed, the horizontal leaves in a stem-and-leaf plot correspond to the vertical bars of a histogram.

Stem-and-leaf plots can be used for decimal numbers also. They are produced by using the integer portion of the data as the stem and the digit after the decimal point as the leaf. If the decimal portion has two figures, the analyst rounds the last digit so only one decimal place remains. The following shows a stem-and-leaf plot using whole numbers:

```
5: 3
5: 4 4
5: 5 5 5
5: 6 6 6 6 6
5: 7 7 7 7 7 7 7 7
5: 8 8 8 8 8 8 8 8 8
5: 9 9 9 9 9 9 9 9 9 9 9
6: 0 0 0 0 0 0 0 0
6: 1 1 1 1 1 1 1
6: 2 2 2 2 2
6: 3 3 3
6: 4 4
6: 5
```

And the following is a stem-and-leaf plot using decimal numbers:

```
5: 3
6: 3 4
7: 4 5 6 7
8: 2 6 7 8 8
9: 2 3 4 5 5 6 8 9
10: 1 1 2 3 5 5 6 8 8 9
11: 0 1 2 3 3 4 5 5 6 7 7 9
12: 1 2 3 3 4 4 5 6 8
13: 1 1 2 3 5 9
14: 2 4 7 9
15: 4 3 7
16: 5 7
17: 1
```

Note that the length of the leaves shown in the leaf-and-stem plot for whole numbers corresponds to the frequencies in table 17.2 when represented in a frequency table. To construct a frequency table, the analyst arranges the data by categories, increments, or—as is the case with table 17.2—groups.

KEY WORDS AND PHRASES

The key words and phrases associated with this chapter are listed below. Demonstrate your understanding of each by writing a short definition or explanation in one or two sentences.

- Bar chart;
- Chart;
- Figure;
- Frequency polygon;
- Frequency table;
- Graph;
- Histogram;
- Line chart;
- Pictograph;
- Pie chart;
- Stem-and-leaf plot;
- Table;
- *X* axis; and
- *Y* axis.

STUDY QUESTIONS

1. Explain why an analyst would use graphs to display statistical results in a written report or a briefing aided by electronic slides.
2. What are the types of visual representations available for intelligence analysts to use? List a few.

Table 17.2.   An Example of a Frequency Table

Groups	Frequency
51–55	6
56–60	44
61–65	18

3. Discuss the difference between a histogram and a frequency polygon.
4. Describe the attributes of a correctly laid-out table of statistical data.
5. When would an analyst use a stem-and-leaf plot? Give an example.

LEARNING ACTIVITY

Suppose an intelligence analyst is preparing a briefing to industry executives of the occurrences of trademark violations in a major city known for its high international tourist traffic. Over the past year, field operatives have noted the following numbers of individuals trading in counterfeit trademarked goods for each month starting in January and ending in December: 27, 35, 17, 13, 42, 37, 52, 48, 20, 24, 12, and 30. Select an appropriate method of displaying these data for an electronic slide presentation, and prepare the slide using a commercial software package.

NOTE

1. Henry Prunckun, "Operation El Dorado Canyon: A Military Solution to the Law Enforcement Problem of Terrorism—A Quantitative Analysis" (master's thesis, University of South Australia, 1994), 52.

18

❦

Threat, Vulnerability, and Risk Assessments

This chapter focuses on analytic techniques in relation to the dangers posed by terrorists to personnel and critical infrastructure. The methods of analysis discussed include:

1. Threat analysis;
2. Vulnerability analysis;
3. Risk analysis; and
4. Prevention, preparation, response, and recovery planning.

INTRODUCTION

The term *counterterrorism* is in widespread use and describes defensive countermeasures. For some reason it has been used instead of the term *antiterrorism*, which strictly speaking, is the correct term for this activity. Counterterrorism involves offensive measures taken to "prevent, deter, pre-empt, and respond to terrorism,"[1] so the more accurate term should be *antiterrorism*.

Nevertheless, the term *counterterrorism* is used in this text because it is the most common term used by law enforcement and security agencies worldwide. Notwithstanding, the distinction between these two forms of activity should be kept in mind whenever discussing approaches to deal with terrorists.

BACKGROUND

Threat analysis is the first of three integrated phases in developing a counterterrorism plan.[2] The two subsequent phases are vulnerability

283

analysis and risk analysis. The results of these three pieces of analytic work lay the groundwork for crafting a policy that addresses *prevention, preparation, response,* and *recovery* (PPRR). In other words, all of the techniques contained within this chapter are intrinsically linked and act as building blocks to form a comprehensive methodology for counterterrorism analysis.

These steps in summary form are:

1. Identify the threat(s);
2. Explore vulnerabilities to this threat(s);
3. Gauge the likelihood that the threat(s) will eventuate;
4. Assess the consequence the threat will have; and
5. Construct a PPRR plan.

Consider the following example of how these steps are applied in practice:

1. Threat—cyber attack via an e-mail-borne virus;
2. Vulnerability—the agency's servers and workstations via the Internet;
3. Likelihood—greater than 85 percent probability;
4. Consequence—moderate to severe loss of computer resources; and
5. PPRR—develop a plan that does four things: attempts to prevent such an attack (prevention); prepares the agency for such an attack if prevention measures fail (preparation); guides the agency in the actions it needs to take to respond to an attack that is under way or has occurred (response); and suggests what needs to be done to aid the agency in recovery once an attack has passed (recovery).

Although discussed here as a packaged approach to counterterrorism, any one of these analyses can be carried out on its own or applied to problems other than terrorism. For instance, a threat analysis could be carried out in relation to gangs, and a risk assessment could be conducted in relation to a person or group acting criminally.

THREAT ANALYSIS

The purpose of carrying out a threat analysis is to identify problems that personnel and physical assets may face.[3] A *threat* is a person's resolve to inflict harm on another. It is important to note that a threat cannot be posed by a force of nature or natural event—these are *hazards*. Only people can pose a threat, as they need *intent* and *capability* (or organizations, associations, businesses, or other forms of body corporates because they are staffed by people), which will be explained shortly.

Threats can be made against most entities—people, organizations, and nations—and this is done by a *threat agent*. The potential harm can be in many forms and can be suffered either physically or emotionally/mentally. Threat agents do not have to openly declare their resolve to cause harm in order to constitute a threat, though explicit words or actions make it easier for field operatives to identify the threat agent and for analysts to assess the threat.

Threat analysis acknowledges two key factors—that there needs to be a threat agent (which could be anything from a person to a group of people, or a body corporate/organization) and an object of the threat (i.e., the target—which does not have to be a material target such as a shopping mall or an individual; it can be intangible, such as the threat to national security or the security of a particular venue or event). Stated another way, a threat agent who has intent and capability must be able to harm something. By way of example, a threat agent could be a drug trafficker who is intent on and capable of illegally importing, say, heroin or a group of insurgents who have an intent and capability to destroy a bridge, and so on.

When analysts assess a threat agent, they are gauging whether the agent has *intent* and *capability* to produce harm to a target (this is why naturally occurring events cannot be threats). To weigh whether the agent has intent and capability, analysts need to establish two elements for each of these factors: *desire* and *expectation* (or *ability*) for intent, and *knowledge* and *resources* for capability. These considerations are shown diagrammatically in figure 18.1. As an equation, threat is expressed as:

$$(desire + expectation) + (knowledge + resources) = threat$$

Desire can be described as the threat agents' enthusiasm to cause harm in pursuit of their goal. Expectation is the confidence the threat agents have that they will achieve their goal if their plan is carried out. Knowledge is having information that will allow the threat agents to use or construct devices or carry out processes that are necessary for achieving their goal. Resources include skills (or experience) and materials needed to act on their plan.

Two additional concepts that are worth noting are:
threat intent = the optimism a threat agent has about successfully attacking a target; and *threat capability = the force a threat agent can bring to bear on a target.*

A fishbone analysis could be used to show the factors that contribute to each of these elements in a cause-and-effect relationship (refer to figure 11.2 in chapter 11). Recall from our discussion of fishbone analysis in chapter 11 that at the fish's head is listed the problem to be investi-gated—in this case, it is "threat." The major bones of the fish constitute the important categories of information concerning the problem—desire, expectation, knowledge, and resources. From each of these major bones, minor bones sprout, comprising the contributing factors that constitute each of the four categories.

Analysts need to consider the context of the threat, their agency's mis-sion, and the list of potential targets when adopting a model in order to aid them in determining the threat environment. A generic model for cal-culating threats might look something like the example of a threat sum-mary depicted in table 18.1. This model can be modified to suit the specific requirements of individual agencies or research projects.

Though models do not eliminate subjectivity, using a model forces the analysts to be transparent about how they calculate threat and, in doing so, they are able to defend their conclusions. You will note that there is no weighting attached to what constitutes, say, an acute or high level of intent. That is because one cannot say how many media announcements it would take from a group like al-Qaeda to constitute such a level of intent. Ideally, some form of conditioning statement would be attached to each of these scale categories so that the decision maker knows what is meant

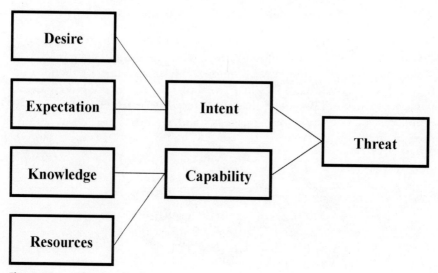

Figure 18.1.   Threat analysis

Table 18.1. Threat Posed to the Orrenabad Community by the Omen Martyrs Faction

Scale	Scores	Tally
Desire		
Negligible	1	
Minimum	2	
Medium	3	3
High	4	
Acute	5	
Expectation		
Negligible	1	
Minimum	2	
Medium	3	3
High	4	
Acute	5	
Total Intent		**6**
Knowledge		
Negligible	1	
Minimum	2	
Medium	3	3
High	4	
Acute	5	
Resources		
Negligible	1	
Minimum	2	
Medium	3	3
High	4	
Acute	5	
Total Capability		**6**
Threat Coefficient		**12**

by high intent, low intent, and so forth. An example of how such a conditioning-statement scale could be constructed is shown later in this chapter, in table 18.5 (note: constructing such a conditioning-statement scale for this table forms one of the learning activities at the end of this chapter).

In addition, models do not eliminate miscalculations because of inadvertent skewing. Note in table 18.1 that intent is calculated by adding desire with expectation, and in turn, this sum is added to the sum of knowledge and resources (and will range from a low of 4 to the maximum of 20). The process of adding limits the spread of values, whereas the process of multiplying any of these scores would increase the values.

For instance, if all scores were multiplied—that is, substituting multiplication for addition—as per the equation, the range would be spread from 1 to 625.

The precision of this wide range of values diminishes the analyst's ability to accurately determine either intent or capability. Therefore, it is suggested that adding all values, rather than multiplying them, will reduce the spread and, therefore, maintain the threat coefficient as an *indicator* rather than promote it as a reflection of its absolute condition. Even if the analyst multiplied desire and expectation and knowledge and resources but added the resulting sums, it would still yield a very wide spread—from 2 to 50—as would the opposite, that is, multiplying the sums that comprise intent and capability, from 4 to 100.

Having said that, two additional issues need to be noted: (1) there is still a need to provide conditioning statements so that the reader of the intelligence report understands what is meant by a medium threat intent and capability (e.g., along the lines of table 18.5); and (2) "unknowns" are not accommodated in this model. Analysts should always be cognizant of unknowns in the form of what are termed *black swans*[4]—unexpected and unforeseeable events, or in this context, threat agents.

The threat coefficient obtained from this analysis is then compared against a reference table to gauge where it sits on the continuum of danger of attack. The scale suggested in table 18.2 can be varied with additional qualifiers, or it can be collapsed if the number is deemed too many. Likewise, how the incremental breakdown of coefficients is determined will depend on whether the agency is willing to accept the risk that a threat agent may slip under its gaze by raising the categories of negligible and minimum. In the end, the number and their descriptors need to make sense in the context of the asset being protected. That is, each of the descriptors needs to have a conditioning statement attached to it to define what is meant by negligible, minimum, medium, high, and acute. Table 18.5 (shown later in this chapter) is an example.

Threats are context dependent, and what forms a threat in a business setting does not necessarily form a threat in a military setting or national

Table 18.2. Example of a Threat Coefficient Scale

Threat Level	Coefficient
Negligible	4–6
Minimum	7–10
Medium	11–15
High	16–18
Acute	19–20

security setting (though the opposite may be true). Bearing this in mind, an example from the military will be discussed to illustrate the threat analysis method. In a low-intensity conflict, threats can range from spontaneous street demonstrations by the local population at one end to terrorist bombings and confrontations with insurgent or guerrilla units at the other end. The techniques for assessing the elements of a threat can vary depending on the issue under investigation and the analyst's personal preference or the agency's policy.

Nevertheless, the approach is to weight each element using some verifiable means that is open to third-party scrutiny. For instance, an analyst may use a force field analysis to judge whether there are threats in Country Q associated with a low-intensity campaign being prosecuted by friendly military units. Likewise, the nominal group technique could be employed not only to assess the four elements of a threat (i.e., desire, expectation, knowledge, and resources) but also to generate a list of possible threat agents (i.e., belligerents) to compare the elements against each other. Participants for such a group could be drawn from subject experts or operational specialists or a mixture of both. Some of the other analytic techniques discussed in chapter 11 can also be used.

There is no firm rule on how this analysis should be done. One way of contextualizing threats is to see them as *threat communities*. Some examples of threat communities pertaining to malicious human threats include:

External

- Competitors;
- Common thefts;
- Conspiracy theorists and other advocates of pseudoscience beliefs;
- Local gangs;
- Organized criminal groups;
- International or transnational terrorists;
- Domestic terrorists (including offshoots);
- Insurgents and guerrillas;
- Anarchists;
- Domestic anarchists;
- Cyber criminals and cyber vandals;
- Radical political groups;
- Rights campaigners;
- Single-issue lobbyists;
- Spies-for-hire (i.e., former law enforcement, security, military, or intelligence personnel who have turned private operatives); and
- Foreign government intelligence services.

Internal

- Principals of the business or corporation;
- Associates;
- Current employees;
- Former employees;
- Temporary staff; and
- Contractors.

These threat communities can be subdivided into more distinct groups if there is a need—for instance, extremist rights campaigners could be classified into the following extremist subgroups: political extremists, religious extremists, single-issue extremists. But bear in mind that membership in one threat community (or subcommunity) does not exclude that person being a member of another or several other threat communities.

> When compiling a threat profile, targets can and should be considered in terms of their criticality, cost (either as a direct loss or an indirect or consequential loss due to disruption), or sensitivity (e.g., compromised information). This is because targets that do not possess any of these attributes may not be considered by threat agents with the same weight.

To better understand the "who" that comprises a threat community, analysts need to compile a *threat profile*. The profile needs to be adequate (perfection is rarely, if ever, obtainable) in order to understand the threat environment, which aids the next phase in counterterrorism analysis—that is, vulnerability analysis. In the meantime, consider the threat profile shown in table 18.3 as an example that demonstrates the important aspects of a fictitious threat agent (the order can be rearranged to suit the analyst's research project, and other factors can be added if these are deemed inadequate to communicate the message).

Threats—Strategic, Operational, and Tactical

Threats of a strategic nature could be argued to be potential *hazards*, not threats. This is because such issues have not manifested themselves into an expression of intent. Take, for example, a less than friendly country that might be constructing a nuclear facility which is several years away from completion. This construction project, the country explains, will be for peaceful purposes—electric power generation. However, given the

Table 18.3. Threat Profile for the Notional Omen Martyrs Faction

Summary	Observations
Desire	
Targets	Objects that represent Western values or people who do not ascribe to their interpretation of their faith (including other believers).
Affiliation	Totally autonomous.
Recruitment	Educated local ethnic population.
Target characteristics	Symbolic and iconic objects that afford high visibility and hence high media coverage.
Tactics	Targets mass gathering, critical infrastructure, communications, mass transport, and distribution chains.
Expectation	
Motivation	Radical religious ideology.
Intent	Extensive destruction.
Tolerance to risk	High.
Self-sacrifice	Very accepting.
Willingness to inflict collateral harm	Extreme.
Knowledge	
Planning	Based on target acquisition intelligence through fixed and mobile surveillance, informants.
Information	Open source data collection as well as access to declassified military manuals.
Training	Low-grade, informal facilities. Though training standards are crude, knowledge transfer is effective.
International connections	Training and ideological support.
Resources	
Financing	Extortion and kidnapping the wealthy.
Weapons	Improvised explosives and small arms.
Skills	Attack vector dependent:
	• Computer-based—very low;
	• Electronic/communications—moderate;
	• Small arms—high; and
	• Explosives—very high.

construction design, nuclear subject experts have concluded that the facility and/or the material produced could also be used to construct a nuclear weapon. So in this case it would be correct to term this country's nuclear project a *strategic hazard* rather than a *strategic threat*.

Analysts should note that some scholars use the term *harm* instead of *hazard*. This is not the view adapted by this book because harm is seen as the result (real or potential) of some event or action (*harm* is also referred to as *impact* or *consequence*). Because harm is what has, will, or is likely to occur, it is viewed here as incorrect in this context. *Impact* (i.e., harm) is discussed in the next section on vulnerability analysis, and it is explained why that term is not correct in this context. It is recommended that analysts make themselves aware of the liberal use of these other terms, but use the term *hazard* in intelligence briefings and reports.

The concept of a threat is normally reserved for matters that are either operational or tactical. For instance, an operational threat is an entity, such as an organization that is waging a campaign that is widely spread —like an organized crime group that is involved in the rebirthing of stolen motor vehicles. An example of a tactical threat is an individual or group who is the target of immediate action. Such actions might include prevention, detection, or enforcement. In some cases, actions taken to deal with tactical threats might impact an operational level threat. Using the rebirthing of stolen motor vehicles example, if law enforcers in, say, the New England area raid several individuals who are dismantling cars in their garages in several adjoining states, these actions might impact organized crime in that area (an operational threat).

VULNERABILITY ANALYSIS

Vulnerability can be described as a weakness in an *asset* that can be exploited by a threat agent. The term *asset* is being used in this context to denote a resource that requires protection.[5] A resource can be a person or group of people or a physical entity (e.g., a piece of critical infrastructure). Viewed another way, vulnerability is an asset's capability to withstand harm inflected by a threat. Harm can be anything from experiencing a minor nuisance event to a situation that is catastrophic.

Vulnerability is a function of several factors—attractiveness of the target, feasibility of carrying out an attack, and potential impact (i.e., potential *harm* as discussed in the previous section). This model is shown diagrammatically in figure 18.2. Usually these factors entail such considerations as status of the target, potential for the attack to succeed, potential for the threat agent to get away with the attack, and potential to inflict loss. These factors can be weighed against measures to mitigate loss and to deter or prevent attack on an asset (e.g., through a force field analysis).

Formulae-based analyses are popular among law enforcement and security agencies engaged in counterterrorism, and although these vary from agency to agency, they usually follow a basic stepwise formula. The one below shows the generic approach:

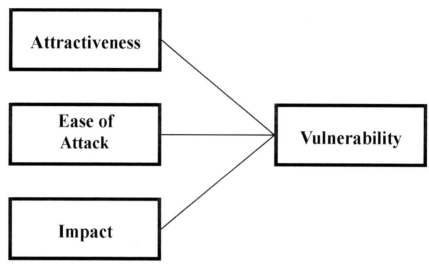

Figure 18.2.   Vulnerability analysis

1. Define what constitutes an asset (critical infrastructure, transport network, food chain, distribution hubs, or any of the essential services—e.g., electric power, gas, potable water, sewerage, etc.);
2. Sort these assets into categories;
3. Assign a grade or level of importance to each asset; and
4. Identify potential impact on the asset if it suffers harm.

As there is no one single criterion for calculating vulnerability because each class of asset may require special considerations to be taken into account (and there may be agency protocols that take precedence also), one general approach is to use a model such as:

target attractiveness + ease of attack + impact = vulnerability

To operationalize *attractiveness*, the analyst could ask questions along the following lines and tabulate the results to insert into the model:

- Is the target readily recognizable? Rather than answer this question in a dichotomous way (i.e., using nominal data—yes/no), the analyst could use ordinal data to give greater precision to the overall vulnerability indicator; for example, is the target recognizable internationally in the same way the World Trade Center was, or is it recognizable only nationally, statewide, or just locally? (See figure 18.3.)
- Is the target the subject of media attention/coverage? Again, an analyst could construct a scale of attention from rarely to frequently/

weekly. Coverage could be in the local press or through to global newscasters.

- Does the target have a symbolic status in terms of historical, cultural, religious, or other importance? The analyst could assess this factor as having no symbolic status to having multiple imports.

Attractiveness needs to be placed in context with the threat agent. For instance, some scholars are of the view that some Islamic extremist groups see assets that represent Western culture or symbolize Western values as attractive.[6] To operationalize the concept *ease of attack*, the analyst could ask these types of questions:

- How difficult would it be for the threat agent to predict the peak attendance times at the target? Establish a scale from certain (as in the case of published opening hours) to very difficult (in the case of a training center located in a remote area and opened only for ad hoc lectures).
- Are there security measures in place (e.g., calculated on a scale of low to high deterrence or low to high prevention)?

Questions that probe the existence and extent of controls (or lack thereof) can also be asked to gauge ease of attack. On the one hand, if

Figure 18.3.   International recognition—the former World Trade Center.
Photograph by author

there is a high degree of control effectiveness, this will usually reduce ease. On the other hand, if there is a low level of control effectiveness, it will increase ease. Analysts should be mindful that, with some targets, even a small reduction in control effectiveness can result in a disproportional increase in ease of attack. *Impact* could be operationalized by questions like:

- What are the numbers of people frequenting the target? Establish a scale ranging from a few daily/weekly/monthly to hundreds or thousands daily/weekly/monthly. Are these same people attracted from the local community, or are they international tourists?
- In dollar terms, what would the financial impact of an attack be if the asset was disrupted, incapacitated, or destroyed? Or it could be put in terms of hours without operation, units of production, and so on.

Impact is predicated on an assumption that terrorists want their attacks to result in large numbers of deaths. This may be true in the al-Qaeda-focused climate that existed at the time of this writing, but such an assumption may not always be valid; for instance, there may exist a nationalist-focused group that seeks to destroy infrastructure rather than kill people. In such a case, these terrorists may view heavy public traffic as an inhibitor to ease of attack. The two paradigms could be described as *effect-based attacks* versus *event-based attacks*.[7]

A template for calculating vulnerability might look something like that shown in table 18.4.

The vulnerability coefficient derived from this analysis is then compared against a reference table to gauge where it sits on the continuum of susceptibility to attack. The scale can be increased with additional qualifiers (i.e., conditioning statements), or it could be collapsed if the number is deemed too many. In the end, the number and the descriptors need to make sense in the context of the asset being protected (the left-hand and center columns of table 18.5). Qualitative descriptors (i.e., conditioning statements) can be added for each category as shown in the right-hand column of table 18.5.

Note that *consequence* is not a factor that is considered in a threat assessment. It is, however, considered in a risk assessment (see the following section).

RISK ANALYSIS

Risk is a function of *likelihood* and *consequence*. The term *probability* is sometimes used instead of *likelihood*. Both are acceptable.

Table 18.4. Vulnerability of the City's Main Bridge over the Orrenabad River

Scale	Scores	Tally
Attractiveness		
Negligible	1	
Minimum	2	
Medium	3	3
High	4	
Acute	5	
Ease of Attack		
Negligible	1	
Minimum	2	
Medium	3	3
High	4	
Acute	5	
Impact		
Negligible	1	
Minimum	2	
Medium	3	3
High	4	
Acute	5	
Vulnerability Coefficient		**9**

A risk assessment can be carried out in relation to almost any situation; it is not just for issues of grave concern. Nor is risk management solely for counterterrorism; risk analysis techniques can be applied to situations that may be the target of criminals or criminal organizations (e.g., gangs) not associated with terrorism. Nevertheless, analyzing risk allows analysts to recommend measures that will provide field commanders with the ability to:

- Accept the risk as-is; or
- Treat the risk (which includes such decisions as to avoid the risk altogether, mitigate the risk, or defer the risk to another person or agency).

In intelligence research, analysts can focus on a wide range of risks. These can vary from the minor—say, a noncritical facility—to risks that liberal democratic nations face from the likes of weak and corrupt government; rogue states; sub-state and trans-state actors; organized criminal, radical ethnic, racial, and religious groups; and ultra-right-wing political groups.

Table 18.5. Examples of Vulnerability Coefficients with Qualifiers (i.e., Conditioning Statements)

Vulnerability	Coefficient	Qualifier (i.e., conditioning statements)
Negligible	1–3	• Can only be attacked successfully if the threat agent has an acute threat coefficient; or • Has little or no importance; or • The range of security measures makes attack very difficult; or • If attacked, the information has little utility to cause harm.
Minimum	4–6	• Can only be attacked successfully if the threat agent has a high coefficient (or greater); or • Has limited importance; or • The range of security measures makes attack difficult; or • If attacked, the information has only some utility to cause harm.
Medium	7–9	• Can only be successfully penetrated if the threat agent has a medium coefficient (or greater); or • Has reasonable amount of importance associated with it; or • The range of security measures makes penetration moderately difficult; or • If attacked, the information has a moderate level of utility to cause harm.
High	10–12	• Can only be successfully attacked if the threat agent has a minimum threat coefficient (or greater); or • Has a sizable amount of importance associated with it; or • The range of security measures makes penetration undemanding; or • If attacked, the information has a high degree of utility to cause harm.
Acute	13–15	• Can only be successfully attacked if the threat agent has a low threat coefficient (or greater); or • Has a very high level of importance associated with it; or • The range of security measures is nonexistent; or • If attacked, the information will cause immediate and/or extreme harm.

Internationally, risk analysis is the subject of a standard. The Swiss-based International Organization for Standardization (ISO) has published a document that puts forward a common approach for dealing with risk by providing generic guidelines in relation to the principles for how risk is managed.[8] In Australia, as well as in New Zealand, uniformity in risk management is specified by AS/NZS 31000:2009. This document is published through a joint venture by these two organizations: Standards Australia and Standards New Zealand. AS/NZ 31000:2009, along with the ISO:31000, can be applied to a number of activities, decisions, or operations in the private and public sectors as well as the military. They can also be applied by nonprofit organizations, community groups, and individuals.

Some of the key terms used in risk management include the technique that is considered here—*risk analysis*—as well as *risk, risk assessment*, and *risk management*. According to AS/NZ 31000:2009, risk is "the effect of uncertainty on objects."[9] Risk assessment is "the overall process of risk analysis and risk evaluation,"[10] and risk management is "the coordinated activities to direct and control an organization with regard to risk."[11]

Understanding these terms helps distinguish the process of managing risk from the analytic process of assessing risk using the equation:

$$risk = likelihood + consequence$$

Likelihood refers to the probability of "a specific event or outcome, measured by a ratio of specific events or outcomes to the total number of possible events or outcomes." *Consequence* is defined as "the outcome of an event affecting objects."[12] Likelihood and consequence are evaluated in the analysis phase of the risk management cycle. This analytic cycle comprises phases shown diagrammatically in figure 18.4.[13] Step by step:

1. Use two tools—in the form of scales—to evaluate a target's risk rating (i.e., the asset under consideration). These two scales consist of a likelihood scale (see table 18.6) and the consequences scale (see table 18.7).
2. The results of these two assessments are then fed into a risk rating matrix (see table 18.8) that returns a risk rating coefficient.
3. Finally, the analyst looks up the risk rating coefficient on the risk evaluation scale (see table 18.9) in order to determine what actions (if any) are required.

In addition to the descriptors listed in the tables 18.6 to 18.9,[14] a set of conditioning statements may be necessary, along the lines of those contained in table 18.5. This also applies to the descriptors contained in table 18.7.

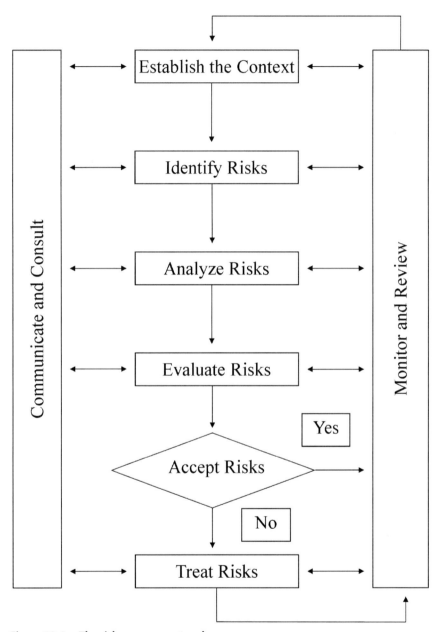

Figure 18.4. The risk management cycle.

An example of low-risk events includes:

- An event that would occur rarely and would result in insignificant consequences (reflected in table 18.8 as E1); or
- An event that is unlikely to occur and would result in minor consequences (reflected in table 18.8 as D2).

Examples of high-risk situations include:

- An event that would occur rarely but would result in catastrophic consequences (reflected in table 18.8 as E5); or
- An event that is likely to occur and have minor consequences (reflected in table 18.8 as B2).

Treating Risk

Once each risk is assessed in this way, they can be positioned on the risk rating matrix (see table 18.8) so they can be compared with each other in order to prioritize treatment options. Take for instance the following events considered by troops stationed in Country Q:

- The risk posed by a person-borne suicide bomb to a public meeting place could be located at C5 (possibly with catastrophic consequences—therefore, it is an extreme risk); or

Table 18.6.   Typical Example of a Likelihood Scale

Rank	Likelihood	Descriptors
A	Almost Certain	The situation is expected to happen
B	Likely	The situation will probably occur
C	Possible	The situation should occur at some time
D	Unlikely	The situation could occur at some time
E	Rare	The situation would only occur under exceptional circumstances

Table 18.7.   Typical Example of a Consequences Scale

Rank	Consequence	Descriptors
1	Insignificant	Will only have a small impact
2	Minor	Will have a minor level of impact
3	Moderate	Will cause considerable impact
4	Major	Will cause noticeable impact
5	Catastrophic	Will cause systems and/or operations to fail with high impact

Table 18.8. Typical Example of a Risk Rating Matrix

	Consequences				
Likelihood	1 Insignificant	2 Minor	3 Moderate	4 Major	5 Catastrophic
A Almost Certain	Moderate	High	Extreme	Extreme	Extreme
B Likely	Moderate	High	High	Extreme	Extreme
C Possible	Low	Moderate	High	Extreme	Extreme
D Unlikely	Low	Low	Moderate	High	Extreme
E Rare	Low	Low	Moderate	High	High

- Violence as a result of a street demonstration could be located at B3 (likely with moderate consequences—so the risk is high).

The scale provided in the risk rating table (see table 18.9) is useful for judging whether the analyst recommends accepting the risk or treating the risk (and, if so, to what extent). Without the risk assessment process, the recommendations of the analyst could be called into question as an overreaction or, equally, deemed an underestimate of the seriousness of the situation. These models curb subjectivity to some extent by providing transparency about how analysts make their calculations.

According to Emergency Management Australia (EMA), some treatment options for critical infrastructure include awareness and vigilance, communication and consultation, engineering options, monitoring and review, resource management, security and surveillance, and community capability and self-reliance.[1]

1. Emergency Management Australia, *Critical Infrastructure Emergency Risk Management and Assurance*, second edition (Canberra, Australia: Attorney General's Department, 2004), 43.

Table 18.9. Typical Example of a Risk Evaluation Scale

Risk Rating and Suggested Actions for Treatment	
Low Risk	Manage using standard operating procedures.
Moderate Risk	Outline specific management actions that need to be taken.
High Risk	Create a business contiguity plan and a response plan (test annually).
Extreme Risk	Urgent actions are necessary (in addition to those per high risk).

Although the risk rating (see table 18.9) shows what is a generally accepted distribution of risk levels,[15] analysts will need to make their own judgments as to where these transition points take place. Many times this will be a topic for discussion with the employing agency or a matter set by policy. But by using a systematic approach to risk management, analysts can reduce the likelihood and lessen consequences through the application of technology, science, or personal or collective effort.

PPRR PLANNING

There are four elements to PPRR policy development—prevention, preparation, response, and recovery. Prevention considers the risk and tries to implement ways that could stop it from happening. Preparedness acknowledges that despite preventative measures, the event may still occur, so one should prepare for it. If it does occur, response is that part of the plan that deals with how agencies will mobilize and take action (and what type of action, etc.).

The final element provides guidance for recovery operation. This aspect of the plan anticipates the worst-case scenario. That is, preventative measures have failed, and preparation measures may have mitigated the impact to some degree, but it still occurred; response has contained and brought the event to an end, but it is now time to recover from the event's effects.

Even though this chapter addresses PPRR from a counterterrorist point of view, it should be borne in mind that when planning for one type of event, it is prudent to consider actions to cover what is termed *all hazards*. For instance, if an analyst is considering the impact of a particular terrorist event, then he or she should consider the event occurring as a result of nature—wildfire, flood, earthquake, and so on.

When compiling a PPRR plan, try to avoid constructing the plan in such a way that elements form either conceptual or real barriers between them—there is usually no clear delineation between the elements, though they may be expressed in these terms. Also, bear in mind that not each element will carry the same weight of importance—the four elements may not be equal. In fact, some elements may not have any strategies or treatments, or few, or minimal.

Further, although the elements are cited in a sequence—PPRR—they may be put into action at the same time; for instance, response and recovery can, and in most cases, should start at the same time, as they are inextricably linked (see figure 18.5). As it is arguable that until recovery starts, the target of attack cannot function, recovery should be considered at the earliest opportunity.

Finally, though the language appears to contain action-oriented terms, the treatments do not have to be physically based options. Options involving social dimensions are also needed as, arguably, people are the ultimate targets of a terrorist attack (recall the basic philosophy of the terrorist: *kill one, frighten ten thousand*). Analysts should try to keep their thinking about treatments broad and innovative.

KEY WORDS AND PHRASES

The key words and phrases associated with this chapter are listed below. Demonstrate your understanding of each by writing a short definition or explanation in one or two sentences.

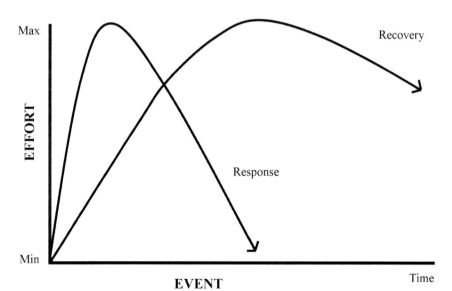

Figure 18.5. Comparison of response and recovery efforts.

- All hazards;
- Attractiveness;
- Capability;
- Coefficient;
- Consequence;
- Desire;
- Ease of attack;
- Expectation;
- Impact;
- Intent;
- Knowledge;
- Likelihood;
- PPRR;
- Resources;
- Risk;
- Threat;
- Threat agent;
- Threat communities;
- Threat profile;
- Treatment; and
- Vulnerability.

STUDY QUESTIONS

1. List the elements that comprise a threat analysis. Describe each and explain why each is important to understanding a threat.
2. List the elements that comprise a vulnerability analysis. Describe each and explain why each is important to understanding the concept of vulnerability.
3. List the elements that comprise a risk analysis. Describe each and explain why each is important to the understanding of risk.
4. List the elements that comprise a PPRR plan. Describe each and explain why each is important to counterterrorism.

LEARNING ACTIVITY

1. Suppose the agency in your jurisdiction that is responsible for monitoring threats of subversion and terrorism has assessed that the water pipeline that connects the city's drinking supply (say, a reservoir of some description) to your city's population is vulnerable to

attack by the notional Omen Martyrs Faction. Using PPRR, devise a plan that considers each of the elements.

2. Using table 18.5 as an example, construct a set of condition statements that could accompany each of the categories in table 18.1 so that the decision maker knows what is meant by high intent, low intent, and so forth.

NOTES

1. U.S. Department of Defense, *Joint Publication 1-02, Department of Defense Dictionary of Military and Associated Terms* (Washington, DC: Department of Defense, October 18, 2008), 132.

2. These methods can also be adapted for dealing with violent gangs. See Hank Prunckun, ed., *Intelligence and Private Investigation: Developing Sophisticated Methods for Conducting Inquiries* (Springfield, IL: Charles C Thomas, 2013), chapter 8.

3. James F. Broder and Gene Tucker, *Risk and the Security Survey*, fourth edition (Waltham, MA: Butterworth-Heinemann, 2012), 316.

4. See, for example, a discussion of such unknowns in a number of contexts: Nassim Nicholas Taleb, *The Black Swan: The Impact of the Highly Improbable* (New York: Random House, 2007).

5. See for instance, Mary Lynn Garcia, *Vulnerability Assessment of Physical Protection Systems* (Burlington, MA: Elsevier Butterworth-Heinemann, 2006).

6. Carl Hammer, *Tide of Terror: America, Islamic Extremism, and the War on Terror* (Boulder, CO: Paladin Press, 2003).

7. Dr. Victoria Herrington, Australian Graduate School of Policing and Security, Sydney, personal communication, May 3, 2009.

8. International Organization for Standardization, *ISO 31000: Risk Management—Guidelines on Principles and Implementation of Risk Management* (Geneva, Switzerland: ISO, 2009).

9. International Organization for Standardization, 2009, 1.

10. International Organization for Standardization, 2009, 4.

11. International Organization for Standardization, 2009, 2.

12. International Organization for Standardization, 2009, 5.

13. International Organization for Standardization, 2009, 14.

14. Queensland Government and Local Government Association, *Local Government Counter-Terrorism Risk Management Kit* (Brisbane, Australia: Queensland Government and Local Government Association, 2004), 16.

15. Queensland Government and Local Government Association, *Local Government Counter-Terrorism Risk Management Kit*, 16.

19

❧

Strategic Intelligence
Assessments

This chapter focuses on writing strategic intelligence assessments
including:

1. Strategic intelligence research;
2. Report types;
3. Writing a strategic assessment;
4. Key parts of a strategic assessment;
5. Figures and tables in the assessment; and
6. Some final thoughts on writing the assessment.

STRATEGIC INTELLIGENCE RESEARCH

Strategic intelligence is considered a higher form of intelligence research
because it provides a comprehensive understanding of the target or the
activity at the center of the investigation. It is related to the tactical assessment discussed in chapter 13, but this type of inquiry provides insights
into future possibilities—implications or ramifications for longer-term
policy. As such, the report that is produced is more detailed than a tactical assessment. It is usually the result of many weeks, months, or even
years of research and analysis. Not surprisingly, the analysts who work
on strategic research projects are often subject experts with extensive
knowledge in particular areas and usually hold advanced degrees.

Because this type of report provides options that will aid planning and
policy development, as well as the allocation of a range of resources
(including, perhaps, human resources who will be placed in harm's way),

the report takes on a different form from the short-form reports that are common in operational and tactical intelligence.

> "Strategic intelligence analysis can be considered a specific form of research that addresses any issue at the level of breadth and detail necessary to describe threats, risks, and opportunities in a way that helps determine programs and policies."[1]
>
> 1. Don McDowell, *Strategic Intelligence: A Handbook for Practitioners, Managers, and Users*, revised edition (Lanham, MD: Scarecrow Press, 2008), 5.

REPORT TYPES

The end point of the intelligence cycle, or what is now being termed the *intelligence process*, is to produce a *report* of some description for dissemination. Often termed an *intelligence product*, the central purpose of the report is to inform the decision maker about the problem, how the analyst approached the inquiry, what was learned, what the information means, and what implications the findings might have for the future. Reports take two forms—written reports and oral briefings. Each may take on a variation, which is characterized by graphics or illustrations. These types of reports are summarized in table 19.1. This chapter will examine the strategic *intelligence assessment* or *intelligence estimate*, or as it is sometimes known, the *long-form report*.

WRITING A STRATEGIC ASSESSMENT

Rare is the analyst who can produce a word-perfect report in one draft. Writing an intelligence report is a formative process, not summative. The

Table 19.1.   Summary of Report Types

Products	Form
Operational Reports, Target Profiles, Tactical Assessments, Reports, Memos, Minutes, Strategic Assessments, Strategic Studies, and National Estimates	Written
Briefings	Oral
Charts, Overlays, and Situational Maps	Graphic/Illustration

analyst should not be discouraged if it takes several passes at redrafting before arriving at the final version. During the process, the analyst should seek feedback and advice from colleagues—doing so can only strengthen the document's message.

The objective should be an error-free report with no awkwardness that will distract the reader from the content. A good approach is to read the text through the eyes of the intended audience. Will they understand what is being said? Can they follow the thought sequence? Is it well laid out, and is the organization logical? The length and format of reports vary according to the agency, the type of information the audience wants to know, and the message that is trying to be conveyed.[1]

On the point of academic rigor, analysts should ask themselves: does the intelligence assessment reflect sound research and analysis? The research methods and the analytic techniques used need to be thoughtfully considered as to their appropriateness for the research question or hypothesis. Put another way: will the assessment and its findings withstand critical appraisal by peers? Reading widely and having a substantial personal library that contains texts on applied research methodologies are keys to being successful as a scholar-spy.

"Precision is a hallmark of the intelligence profession. The term itself is synonymous with accuracy and exactness. Say precisely what you mean. Check your facts to be sure they are facts, and if possible, that you have evidence from more than one source."[2]

2. James S. Major, *Writing Classified and Unclassified Papers for National Security*, (Lanham, MD: Scarecrow Press, 2009), 8.

KEY PARTS OF A STRATEGIC ASSESSMENT

Although intelligence agencies are likely to have their own house style for writing strategic intelligence reports, below is a generic format that will give new analysts an idea of what might be contained in strategic assessments or estimates. The length of the various chapters and sections is likely to vary depending on the target/subject/topic, urgency of the report, and any production particulars set down by the intelligence manager. For instance, some chapters, such as the methodology, may be condensed to a few paragraphs or a page, or eliminated altogether. Some sections may be omitted altogether. Nevertheless, this example shows what strategic reports could include and what information would be expected within the individual parts.

Title

The purpose of the assessment is to inform decision makers who authorized the intelligence research project, or key personnel who receive such reports as a matter of course (e.g., routine briefings). Therefore, the assessment's title should capture the reader's attention but, at the same time, accurately reflect the essence of what's contained in the report without sensationalizing the matter. It should not be a long-winded description of the project. A title is short and concise.

Executive Summary

An *executive summary* is like an *abstract* in an academic study and serves the same purpose—it is a summary that provides the reader with an overview of the study's purpose and findings. It appears at the start of the report, before the introduction and background. In the open source literature, indexing services will often use this information as a way of cataloging the report for other researchers to find.

Table of Contents

A table showing the major chapter titles and the major headings within each of the chapters makes for a convenient way to help the reader find information. If subheadings and other minor headings are incorporated into the table, the added details may make the presentation "busy" and therefore confusing to read. This is true in cases where reports total hundreds of pages. The number of headings and minor headings can therefore fill several pages of a table of contents. However, if it is a shorter report, and the inclusion of minor headings in the table will not take the presentation beyond a single page, it might be worth including them.

List of Tables and Figures

If the tables and figures contained in the report are numerous, this section can be separated into two—a list of tables and a list of figures.

Glossary/List of Acronyms

This is a helpful addition for readers who are not familiar with intelligence or industry-related terms. Even if they are, it is worth including this section, as it helps avoid confusion—for instance, the abbreviation *CI* could mean *counterintelligence, competitor intelligence,* or *critical infrastructure.*

Introduction

This chapter contains several sections that lead the reader through the different aspects of the issue under investigation—from the general to the specific.

Background

This is a short, concise section that sets the context of the issue under study by providing an account of the issue surrounding the events or circumstances. It needs to contain enough information so the reader understands the main or central ideas about the issue. It should not repeat what will appear in the literature review (below), but it sets the scene for the more detailed discussion that will appear there.

Rationale

The rationale section presents the reader the motivation, reasoning, or justification for undertaking the research. That is, why was it considered important to conduct the study into the events/situation that were described in the background section? What were the expected benefits of the research? The material in this section should flow logically from the previous section and not repeat the information that appears in the background.

Theoretical Base (optional)

Even though intelligence research is applied research with a practical outcome, there is usually some theory tied to it. If this is the case, this section therefore lays out the theoretical base on which the study was grounded. The analyst needs to outline the theory, its assumptions, and so on, and how it relates to the study. This section leads the reader into the next section, which states the research question. This section needs to be concise and to avoid overly complicated explanations or discussions on the theory—just state it in simple terms so that the reader knows how the research question relates. Being concise is the key here—expansion of the theory and its application can be done, if needed, as part of the literature review.

Research Question or Statement of Guiding Purpose

Having presented the reader with the background of the issue under investigation and the reasoning for undertaking the study, the analyst now needs to state what the research question will be. State the research question in terms of, for instance, "the purpose of the proposal study is to . . . " or "The matter under investigation can be stated as follows . . . ." If the analyst is testing a hypothesis, it can be stated here also.

As the research question is the study's "compass," it needs to be clear and precise. Do not be tempted to explain the methodology or repeat the importance of doing the study (i.e., rationale); just state the research question. One paragraph of about sixty words will do. Note that a research question is just that—it is a question, not many; nor is it a series of related aims or goals. Having a short lead-in to the question is fine, but then state the question, perhaps like this: "In summary, the matter under investigation can be stated as follows: An increase of police on the street will lead to a reduction in victim-reported crime."

Dissemination (optional)

This section may or may not feature in the final assessment depending on the agency's in-house report style. However, dissemination should be considered, as this is the object of the strategic intelligence project. So considering the intended audience is important. To some degree, the analyst should have discussed this in the section on rationale, so this could be incorporated into that section if desired. But dissemination needs to be spelled out—will the results be circulated directly to decision makers under some security caveat, or will it be a highly classified report for operational commanders who may use it to initiate a, say, counterintelligence investigation, or to guide/focus an existing covert or military operation? This section, if not incorporated into the rationale, needs to be short but clear—who is the readership?

Literature Review or Background (optional)

This chapter can consist of one main section or several sections, depending on the complexity of the study. The literature review is the section where the analyst "tells the story" about the issue under investigation.

Even though information is collected, this should not be confused with the data collection phase of the study-in-chief. The literature review simply provides an overview of the issue. In this regard it provides the context for the research, and in a sense this section could be referred to as *background* rather than literature review.

So this section should provide a summary (i.e., in proportion to the overall length of the intelligence assessment) of the research that has been conducted to date and where your proposed study fits into this picture (i.e., the "gap" that this research has fulfilled). But having said that, it should not be an annotated bibliography—the analyst needs to synthesize the key literature in the field, define the variables, and explain how these variables were operationalized and what theory was used to test the hypothesis or explore the research question. This section can also contain a discussion about the study's theoretical base (if used) and the presumed relationship between the variables.

A good way to begin writing this section is to mind-map the concepts related to the issue; then arrange these concepts into a series of headings that form a logical order that "tells the story." The end of the section should lead the reader into the method section.

Methodology (all or parts optional)

This is an important chapter of the intelligence assessment. If the analyst has carefully crafted the study's research question and placed it in its theoretical framework, it will guide the reader through the overall design, providing understanding of what was done and why. Designs discussed in chapter 5 include:

- Evaluation (to plan intervention programs/operations);
- Case study (what is going on);
- Longitudinal study (has there been any change over time?);
- Comparison (are A and B different?);
- Cross-sectional (are A and B different at this point in time?);
- Longitudinal comparison (are A and B different over time?);
- Experimental and quasi-experimental (what effect does A have on B?); or
- A combination.

In this chapter the analyst needs to define the concepts he or she will study so that they can be operationalized and observed, and then measured. The analyst also needs to identify what data will be required (whether these data will be from primary or secondary sources; whether they will be qualitative, quantitative, or both; if the data will come from open sources, empirical observation, or via covert sources, etc.) and how these data will be collated and analyzed (e.g., statistically or content analysis) to test the hypothesis.

The analyst needs to turn his or her mind to the related issues of sample size, how confounding variables will be controlled (i.e., the potential that observations are due to something other than what is being measured), and what limitations these extraneous influences might present for the research (e.g., possible alternative explanations for the relationship between A and B) or the limits inherent in the data (e.g., missing or incomplete data), and so forth.

Data Collection (optional)

In this section the analyst needs to think carefully about what data will be needed in order to answer the research question (or hypothesis) and how these data will be collected. Because what is being described in this section is based on the scientific method of inquiry, it needs to be transparent, with enough detail to allow the reader to replicate the study, should this be desired. Even if the study is not replicated (few intelligence studies ever are), having a transparent and potentially replicable method allows the reader to critically evaluate the robustness of the method. In academia this is known as peer review, and it is the highest form of assessment for research, whether it is secret or not.

To start, a data collection plan should appear in this section (e.g., it could be in the form of a stepwise narrative or outline). The following points are suggestions to consider when writing the data collection section:

1. What data will answer the research question, where are these data held, and what is the best way of obtaining them? In describing these aspects, explain to the reader why the study has concluded that, say, primary data (from sources such as observations or informants/ agents) or secondary data (from open sources) was chosen.
2. What are the theoretical roots for the decisions that were made in (1) above (they cannot be the analyst's opinion—references to scholarly authority is required to support these decisions)?
3. What is considered the best methodological framework to undertake the study (e.g., quantitative, qualitative, or mixed methods), and what is the scholarly thinking that supports this choice (i.e., the analyst will need to support this with references)?
4. If the study is using samples, the analyst will need to ground this in theory that supports such aspects as the sample frame and sample size (e.g., via references to the literature on methodologies).

Data Collation and Analysis (optional)

In this section of the methodology the analyst needs to reflect on the type of data they have decided to collect so that he or she can explain how it was collated and then analyzed. Like the stepwise data collection plan, the analyst needs to explain to the reader how the analytic methods that were used produced the results that were reported (i.e., how these analyses answered the research question). For instance, if the study looked for correlations or relationships between variables, what statistical test(s) were used? If the study examined themes or patterns in unstructured data, what methods were used? Remember, the decisions that were made to use certain processes and/or methods in the study need to be supported with references to the theoretical literature—they cannot be just the opinions of the analyst.

Limitations (optional)

If any major limitations were noted, they need to be discussed in this section. It is important that the procedures and methods discussed here link directly to how the study went about answering the research question or hypothesis. The reader needs to be able to understand what was done (i.e., transparency) in order to be able to replicate it. However, the analyst does not need to list every single possible limitation—just list the major limitations. This is done in order to demonstrate that these shortcomings were taken into consideration when the conclusions were drawn.

Legal Authority (or Ethical Considerations) (optional)

If the research being undertaken is classified, then authorization is presumably based upon some legal framework. This legal foundation thus grants the analyst the authority he or she needs to conduct the study. If that is the case, then the analyst should state the legal basis for the study's authorization in this section.

Contrast this with research in the social and behavioral sciences, where, if the study involves human subjects, then approval to conduct research has to be granted by an ethics committee. But as we are discussing secret intelligence research—not research that will be disseminated via public avenues (e.g., scholarly journals)—the issue of ethics takes on a different dimension.

Results

The *results* chapter can also be referred to as the study's *findings*. Some scholars even refer to this chapter as *analysis*. Any of these terms are fine, as all are in common usage.

Generally speaking, the results chapter of a strategic intelligence report will be presented in three sections. The first is a description of how the data were prepared for analysis. In doing so the analyst needs to be brief so as to focus on the more unique aspects of the analytic technique.

This is then followed by descriptive statistics of the most relevant or important information (if the study is quantitative). But again, these results need to be brief to avoid overwhelming the reader with volumes of results—doing so will run the risk of confusing them by inadvertently obscuring the assessment's central line of reasoning (i.e., the idiom "miss the forest for the trees" applies in this situation). These descriptive data need to be carefully considered and well organized by such techniques as summary tables and graphs, maps, or other diagrams.

The third section links any inferential analyses to the research question or hypothesis. If statistical analyses were not used, then this section would discuss the results of any of the qualitative analytic techniques that were employed to critically analyze the research question, such as a SWOT, PESTO, or force field analysis, or other method for examining unstructured data.

The main consideration here is that only the results are presented. Analysts need to restrain their urge to add their interpretation to these findings—that is done in the next chapter—Discussion.

Discussion

This chapter presents a discussion about the implications or ramifications of the study's results—hence the name *discussion*. The analyst interprets the findings and discusses them in the context of the problem (i.e., background and rationale) and the theoretical base, but always in terms of what it means for the research question. In doing so, caution should be exercised with regard to the language used and the conclusions drawn—journalist terms and phrases should be avoided; that is, language should be objective and conclusions based on the results of the study (i.e., *evidence based*).

Conclusion and Recommendations

Put the intelligence assessment's conclusions or key judgments into a few paragraphs or a page or two, depending on the length of the report. This chapter is intended to reinforce why the research was important and what the ramifications might mean for policy or field operations.

In intelligence writing, there is a need to present findings in tentative terms and to avoid the temptation of absolutism.

References Cited

List the references that have been used in compiling the assessment using the Harvard style of referencing. The purpose of this section is to make transparent the information and intellectual authority the study relied on so the reader can make a judgment as to their scholastic adequacy. Once complete, check the document for consistency in referencing and double-check the report for inadvertent errors, such as not citing other scholars' intellectual work. If the data are classified, then there is likely to be an agency recommendation for how confidential sources are noted; analysts need to adhere to this, as exposure of a covert source or clandestine means of information gathering could risk lives or jeopardize operations past, present, and future.

Appendices or Annexures (optional)

Sometimes it is important to attach information that could prove helpful to the reader in understanding how a judgment was made or to show relationships that would be too voluminous to explain in narrative form in the body of the report. This type of information is better appended to a separate chapter and referred to in the body of the report.

FIGURES AND TABLES IN THE ASSESSMENT

Representing complex data in the strategic intelligence assessment will often take the form of a graph, chart, or table—or a picture, such as a photograph, diagram, or drawing. When doing so, the analyst needs to label this symbolic representation correctly so the reader can identify it in the body of the report.

There are just two terms used to do this—figures and tables. Graphs, charts, diagrams, photographs, and any other illustrations are known as *figures*. A *table* is where data are arranged in rows and columns. Both are numbered sequentially as they appear in the text and independent of each other—that is, figures are numbered as a group, and tables are numbered as another group. If there are more than a few of either, it is worth considering a separate section in the front of the report, after the table of contents, entitled "List of Figures and Tables." If there are a great number of these, then there could be a separate list of figures and a separate list of tables.

SOME FINAL THOUGHTS ON
WRITING THE ASSESSMENT

Analysts will frequently consider the question of what constitutes a good intelligence assessment at the culmination of their inquiries. Given that there are only two ways analysts communicate with decision makers—orally and in writing—deliberation will invariably center on clarity of expression.

Decision makers are people with busy agendas, tight budgets, and substantial pressures, so an analyst's assessment needs to be concise and precise. If a decision maker cannot find the important information in a report, then, to a large degree, the assessment has failed. Remember, the purpose of intelligence is to provide insight so the best possible decision can be made. This is why the above examples and templates have been described in this section—they have been tested many times by many agencies and by busy managers. Some key points to remember are:

- Double-check that the report's conclusions and recommendations dovetail with the original aim of the inquiry;
- Comply with the agency's in-house style or template for reports;
- Do not selectively omit data items if they do not support a "preferred" position—this is not only unethical but can lead to civil and criminal charges against individuals or the agency if a court or commission of inquiry finds the intelligence investigation did not act in good faith;
- Use a number of analytic techniques to distill the data so that the clearest picture emerges;
- When formulating judgments, make it clear what the limitations are—do not give false or misleading indications;
- Follow the agency's policy on making and couching recommendations; provide options in objective terms, and avoid "rivers of blood" prognoses that manipulate decision makers (doing so is bordering on the unethical);
- Aim to write several drafts and have each version reviewed by a colleague who is senior in years of service. This is because he or she is likely to have experienced many of the pitfalls common in presenting reports and can steer a new analyst around the "holes in the road." It is better that someone close to you critically reviews your work than to have an executive in your agency do it with the potential consequence of getting a "black mark" against your reputation; and
- Always get someone else to proofread your work—it is more likely that he or she will note spelling and typographical errors than you as the author.

KEY WORDS AND PHRASES

The key words and phrases associated with this chapter are listed below. Demonstrate your understanding of each by writing a short definition or explanation in one or two sentences.

- Evidence-based conclusions;
- Figures;
- Intelligence product;
- Oral briefing;
- Research question; and
- Tables.

STUDY QUESTIONS

1. Describe the "hourglass" approach to report writing and its advantages.
2. Explain the major parts of a strategic intelligence assessment.
3. Explain the main attributes of the Harvard style of referencing.

LEARNING ACTIVITY

Suppose you are requested to present the findings of your research into an issue facing your agency. To demonstrate the skills learned in this chapter, construct an electronic slide show that will be the basis of your oral presentation. Use the hourglass approach to do this—that is, start from the general, work to the specific, and then end back at the general, like the shape of an hourglass. For the subject material use the topic of report writing. Do this using no more than eight slides—short and concise.

NOTE

1. For a detailed tutorial on preparing accurately written intelligence reports, see James S. Major, *Communicating with Intelligence: Writing and Briefing in the Intelligence and National Security Communities* (Lanham, MD: Scarecrow Press, 2008), and James S. Major, *Writing Classified and Unclassified Papers for National Security* (Lanham, MD: Scarecrow Press, 2009).

20

꩜

Decision Support Analysis

This topic will discuss five important aspects of decision making:

1. Faulty thinking;
2. Making recommendations;
3. The straw man technique;
4. Pareto efficiency analysis; and
5. Paired-ranked analysis.

FAULTY THINKING

Research shows that when groups of people try to decide on a strategy, tactic, or other course of action, they can run the risk of *groupthink*: decisions based not on the best solution, but on a solution that maintains group harmony.[1] The pattern that often lays the groundwork for groupthink features:

Invincibility—if the group's chair or leader considers the option will work, the group is likely to feel confident that luck will be on their side and backs the option;

Unanimity—if the majority hold a view on what to do, then any dissenting view is seen as being wrong; and

Exclusion—if the members of the group are happy to be part of the collective, then they may be reluctant to express a contrary view, as doing so could exclude them from the group, either physically (being expelled) or emotionally (being ignored or given unfriendly responses).

Groupthink is a situation intelligence analysts should avoid. The decision support analytics discussed in this chapter will help analysts provide decision makers with sound guidance.

321

MAKING RECOMMENDATIONS

Recapping from the previous chapter, *recommendations* stem from the conclusions of a study's results. The results are derived from analysis of data. So decision support analysis is the process where the analyst presents the intelligence consumer with a proposed course of action—a recommendation. The action stems directly from the original research question, or it may stem from an ancillary issue the research has discovered.

At first glance, the process of making recommendations appears straightforward—an action is made in the form of a proposal; say, for instance, "it is recommended that ABC intervene by conducting an XYZ-style operation." Of course, such a recommendation would be fleshed out to include, perhaps, time frames and other considerations. However, a recommendation such as this would more often than not be found inadequate by a decision maker because it offers no options or choices.

It is rare that decision makers will base their resolves purely on factual evidence. As strange as this must sound, decision makers balance the facts gleaned from an intelligence report with the political imperatives surrounding the issue being studied. Often social, cultural, historical, economic, and other considerations enter into the decision maker's thoughts before embarking on a course of action.[2]

The art of recommendation making is, therefore, another exercise in intelligence research. Some scholars hold the view that the terms *considerations* or *implications* are better descriptors than *recommendations*. This, they argue, is because analysts ". . . may not be aware of the broader implications of a particular course of action, and may not have access to the relevant information about all the possible options."[3] This argument holds some weight because intelligence analysts are not policy makers, nor are they political advisors, but they are, nevertheless, subject experts who can offer a range of options that can potentially address the original research question—from a "do nothing" option (yes, doing nothing is a choice) to a "gold standard" response (whatever that might be in the context of the issue). In between these "bookend" options will usually be two, three, or more options that can be presented on a scale of increasing efficiency and/or effectiveness (i.e., a projection of the potential impact or consequence of the options).

But diverting for a moment to look at the concepts of *considerations* and *implications*, it is worth noting the subtle differences between these terms and that of *recommendations*, which we already discussed. Quarmby and Young[4] pointed out that a *consideration* focuses ". . . a decision-maker's attention on an issue(s) to help select a course of action." They contrast this explanation with a description of an *implication*, which they point out as ". . . one or more consequences or effects [that] follow from the analysis provided—often in general terms given the implications for policy responses, capability development response, operational posture, and so forth."

Returning our attention to recommendations, each option, as it climbs the scale toward the gold standard, will usually carry a heavier price in terms of financials, personnel, material resources, risk, outcome, consequences, time, or other factors. Although it is not always the role of analysts to cost each of these factors, it may be incumbent upon them to at least brief the decision maker as to the relative expenditure required for each option. Nevertheless, analysts who are putting forward recommendations will need to think in terms of the *preferred option*. That is, there is a point where the mission's objectives could be reached without invoking the gold standard. So by having some idea as to how resource intense the various options will be, this perspective in the report is likely to underscore the preferred option as being the most realistic.

The analyst then presents a rank for the options in order of magnitude. This serves as a benchmark for the decision maker to assess the options within the package of options.

Take for instance an intelligence report recommending a surveillance team be tasked to gather additional information about a criminal gang involved in the "rebirthing" of stolen motor vehicles. Based on the research and using deductive reasoning, this conclusion may be considered to be on firm ground. However, the agency may only have one full-time surveillance team, and tasking it to gather information may place an existing surveillance operation into the importation of heroin at risk.

The police commander needs to understand the risks as well as the costs in order to weigh which operation will be supported. The outcome may never be known to the analyst, as the drug operation may be a classified project and the analyst has no need-to-know. Likewise, the final decision should not be seen as a slight on the analyst's work should the recommendation not be acted upon. It is a balancing act on the part of decision makers to allocate resources under their command as best they can. Nonetheless, it is important analysts are aware that they may be called upon to provide options and cost and other resource estimates so the best decision can be made.

Another way to describe how making recommendations fits into the intelligence research is by couching the thinking in a structured way through the use of five Is, though other conceptual frameworks are certainly possible:

- **Intelligence.** Information gathering and analysis in relation to the problem, leading to making recommendations for intervention (options);

- **Intervention.** The presentation of a set of responses or policy options addressing the proximate and/or distal causes of the problem;
- **Implementation.** The action needed to translate the recommended interventions into practice;
- **Involvement.** The enlistment of key partner agency(s) and/or stakeholder(s) to implement the recommended option; and
- **Impact.** The evaluation of the intervention's outputs and outcomes.

THE STRAW MAN TECHNIQUE

One method to help guide decision makers to the preferred option is to use the straw man technique. The way this is done is to draft a proposal or solution (i.e., the straw man) and place it among a range of alternative options (this can be from the gold standard through to the do-nothing option). The analyst then discusses the preferred option by criticizing the shortcomings of the alternative recommendations; that is, by pointing out how parts of the various options do not meet the goal of the operation/ project. It is important that the shortcomings of the alternative options are discussed in relation to the operation/project's goal or aim. The critiquing process should not be steeped in manipulative or emotional language, as this will undermine the credibility of the report.

In this way, the straw man technique is a beneficial process, as it can provide a springboard for exploring other possibilities. Presenting recommendations in this way allows the decision maker to understand the strengths/benefits of the preferred option when contrasted with other possibilities. It can also facilitate some idea generation by the decision maker who may then request modifications to the preferred option based on political considerations that were beyond the privy of the analyst (e.g., social, cultural, historical, economic, or other considerations). Regardless, the straw man approach to making recommendations increases the chances that the decision maker will be more comfortable with and, hence, "own" the final choice.

PARETO EFFICIENCY ANALYSIS

Pareto efficiency analysis is a quantitative way of prioritizing a list of options, such as recommendations. The Pareto principle derives from the 1879 study of wealth and income by Italian economist Vilfredo Pareto. Pareto observed that 80 percent of the "output" is a result of 20 percent of the "input."[5] It has been used in business as a way of "doing more with less." Therefore, when used in an intelligence context, analysts can use this type of analysis to present a short list of options that are likely to maximize results based on the 80/20 principle.

Pareto analysis relies on a rating for each policy or operational option. As such, a ranking score is often used. But whatever the metric, it must make sense in the context of the intelligence project. Take for instance the following list of policy options that was generated by brainstorming; each relates to how a nation could deal with the continuing example of international pirates.

Scores were assigned for each policy option after all of the ideas were written down. The higher the score, the more important the option was considered. Ideally, scores should be discrete, but if two options are considered equal, there is no reason why the same score cannot be assigned to each. Nevertheless, the final prioritized list must make sense to those who will be implementing the recommendations.

In this example, the goal is to take an aggressive stand against international pirates. The six options under consideration with their corresponding ranked score are:

1. Increase surveillance by aircraft (+5);
2. Institute a system of in-country, covert "coast watchers" based in pirate ports (+3);
3. Increase maritime naval patrols (+6);
4. Monitor radio communications frequencies used by pirates (+2);
5. Randomly search vessels of the class used by pirates (+4); and
6. Scuttle all vessels found carrying offensive arms or explosives (+1).

Based on the scores, the ranking would look like this—from highest to lowest:

1. Increase maritime naval patrols (+6);
2. Increase surveillance by aircraft (+5);
3. Randomly search vessels of the class used by pirates (+4);
4. Institute a system of in-country, covert "coast watchers" (+3);
5. Monitor radio communications frequencies used by pirates (+2); and
6. Scuttle all vessels found carrying offensive arms or explosives (+1).

If the 80/20 principle is applied, then the greatest benefit might be achieved by using the top 20 percent of the options. In this case that would be the first option—increase maritime naval patrols—as this option represents a little over 20 percent of six options under consideration (20 percent by tally). This is not to say this option will achieve the desired policy outcome as it complies with the theory. It is merely a guideline in cases where resource constraints are imposed.

Say, for instance, policy makers ask military commanders what option(s) they would use if some of the solutions put forward had to be cut. In such a case, this type of analysis could help narrow the options to the most effective. Obviously, if resources could stretch to other options, it would be of benefit to include those too, but as a general rule, the 80/20 principle provides a basis for identifying what might be the most efficient.

PAIRED-RANKED ANALYSIS

An effective way to help decide on an option—whether that is a solution to a problem, a choice of equipment, implementing a new process or procedure, or a tactical resolution—is to use paired-ranked analysis. Not to be confused with paired-ranked correlation analysis, this decision support method can be done in a few minutes as long as those making the decision are familiar with the various options at the center of the confusion. This type of analysis is ideally suited to comparisons where not all options are equal. It is also suited to situations where a large number of options are under consideration.

A process that compares various features or aspects of different options with others presents two issues. The first is there is a tendency to compare various features of the options separately, not as a whole. This is because there is usually no practical way of comparing a variety of different aspects with other aspects that are different. So paired-ranked analysis views the option as a whole, not what it consists of through the features that comprise the option. The second issue is that each feature of the different options is given equal importance when trying to weigh them up, when these features do not hold the same standing.

Step by Step

The first step in the process is to note all the options. This can be done in list form manually on paper or electronically on a computer projector if the method is being used in a group decision situation.

Next, the analyst (or group) needs to put themselves in a frame of mind to make several decisions in a straightforward manner.

From the list, the first two options are considered. This simple question is asked—if I had to select option A or B, which would I select? There should be no debate or discussion. If this is done by the analysts alone, then a tick is placed next to the option to start a tally. If this is done in a group situation (say, a group of independent judges), then the option with the most votes gets the tick. Judges should be discouraged from speaking for or against the options, or explaining something about one or both of the options. The various options should have been considered beforehand by all who are participating, thus allowing for an efficient decision based on a series of quick pair-wise comparisons.

Once the first two are compared as an entity, the next option is compared—A with C. Then A with D, A with E, and so on until all options are compared with A.

Then B is compared with C, B with D, and B with E, until this option has exhausted all comparisons.

The analyst continues through all the options until all possible combinations have been made.

Having compared each option against every other option, the analyst inspects the tally. A re-creation of such an analysis relating to a decision-making process for the selection of two-way radio equipment for a covert operation is shown in figure 20.1. This figure highlights that radio D was the favored choice with a tally of three "ticks." This was followed by radio B with two, and then radio C with one. Radio A had no ticks against it. So in this sense, the analysis not only presents the preferred option, but ranks the other options for transparency.

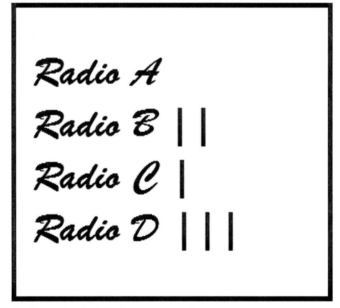

Figure 20.1. Example of a tally for various options.

The process can be conducted in a matter of minutes. Even with long lists of options, it is a quick procedure. In fact, with longer lists the difference between the options is likely to be more pronounced.

Paired-ranked analysis is a simple but effective method for helping people make decisions. In a world where there are often numerous choices, it is difficult to decide on just one—whatever that may be. The paired-ranked analysis forces analysts to decide about each and every combination on the basis of a holistic comparison; that is, they compare one item with another in its entirety. It prevents analysts from engaging in the microanalysis that often leads to confusion. Some may find the forced comparisons a bit unsettling because it is a structured way of thinking. But if used a few times, it is easy to see the clarity it delivers for multiple-choice decisions.

KEY WORDS AND PHRASES

The key words and phrases associated with this chapter are listed below. Demonstrate your understanding of each by writing a short definition or explanation in one or two sentences.

- Consideration;
- Do-nothing option;
- Five Is conceptual framework;
- Gold standard;
- Implication;
- Pareto principle;
- Preferred option;
- Recommendation; and
- Straw man.

STUDY QUESTIONS

1. Why do you think that a do-nothing option is always worth considering? Are there times when doing nothing is better than performing an action? If so, give an example. If you do not think that there are any situations in which doing so might result in a better decision, explain why.
2. Explain why it is important to avoid the use of manipulative or emotional language when engaged in the critiquing process of the straw man technique.

3. Explain the underlying philosophy of the 80/20 rule and give an example that might apply to an issue in your workplace.

LEARNING ACTIVITY

Suppose you have been asked to formulate a recommendation regarding the collection of information for a new strategic intelligence study that will assess the threat that country A poses to country Z's national security (choose two countries that have been in the news recently and who are antagonistic toward each other).

Reflecting on the various forms of information (e.g., primary and secondary) and the various techniques for collecting these data (e.g., open source, unobtrusive, official, confidential, covert, and clandestine, etc.) that have been discussed in previous chapters, formulate a list of recommendations that could be used to gather the information to answer the research question. Include a gold standard option as well as a do-nothing option. Then, using the straw man technique, guide the reader to what you consider to be the preferred option. If you can, try and incorporate the Pareto principle in making this recommendation.

NOTES

1. Irving L. Janis, *Victims of Groupthink: A Psychological Study of Foreign-Policy Decisions and Fiascos* (Boston: Houghton Mifflin, 1972); and Irving L. Janis, *Groupthink: Psychological Studies of Policy Decisions and Fiascos* (Boston: Houghton Mifflin, 1982).

2. Terry-Anne O'Neill, "The Relationship between Intelligence Analysis and Policymaking—Some Issues," *Journal of the Australian Institute of Professional Intelligence Officers* 8, no. 1 (1999): 5–22.

3. Neil Quarmby and Lisa Jane Young, *Managing Intelligence: The Art of Influence* (Annandale, New South Wales: Federation Press, 2010), 32.

4. Neil Quarmby and Lisa Jane Young, *Managing Intelligence: The Art of Influence*, 32.

5. Richard Koch, *The 80/20 Principle: The Secret of Achieving More with Less* (Boston: Nicholas Brealey Publishing, 2007).

21

⑤

Basics of Defensive Counterintelligence

The topics discussed in this chapter focus on defensive counterintelligence—that is, data security and associated issues for those who handle it, including:

1. Information security;
2. Need-to-know and right-to-know;
3. Handling sensitive information;
4. Security clearances; and
5. Storing, protecting, and disposing of intelligence data.

INFORMATION SECURITY

The term *information* reflects all forms of data—ideas, concepts, and plans. *Information security* is therefore concerned with the arrangements for protecting these data once they are recorded. Information security should not be confused with the same term used by the information technology industry, which has a narrow application—that is, the security of data processing hardware and software, and access to the same.

The role information security plays in an intelligence unit is to protect data and reports from *unauthorized* personnel. Keep in mind that the information held by an intelligence unit and the finished intelligence it produces are destined for a very specific client group. These customers have either a *need-to-know* or a *right-to-know*. The delicate subject matter in a unit's files makes it an attractive target to organized criminals, terrorists, corrupt public officials, investigative journalists, and a wide range of antiestablishment ideologues.

The practice of keeping secrets secure is known as *counterintelligence*. Although counterintelligence is comprised of defensive and offensive measures, this chapter introduces the reader to two aspects of defensive counterintelligence—document and personnel security. These topics are essential for anyone working in a security-conscious environment.[1]

NEED-TO-KNOW AND RIGHT-TO-KNOW

In determining the security needs of an intelligence section, we are reminded that it is not absolute security that is sought but a level of security in line with the sensitivity level of the information being guarded. However, having said that, the requirements of security should not eclipse the basic objectives of the unit and impinge upon its operational effectiveness. Security should not inhibit the flow of information or hamper the dissemination of intelligence to a wide range of users. The notion of *need-to-share* will be discussed shortly, but this doctrine states that it is important to share information across the intelligence community as well as within agencies.

The term *need-to-know* is used by the military as well as law enforcement and government agencies to describe sensitive information that needs to be protected by limiting those who receive it. According to the need-to-know doctrine, even if one has a security clearance (e.g., secret), it does not entitle him or her to have access to any or all material classified at the secret level. Before people are given access, they need to demonstrate that they have a need-to-know. In short, this is some justification that access to the report will aid an intelligence project or operational mission.

In the post-9/11 security environment there is also the concept of *need-to-share*. Need-to-share involves a mindset that asks the question "who needs to know this piece of information?" or "who should I be sharing this information with?" Some in the national security arena argue that need-to-know should be replaced with need-to-share. But this may be a simplistic argument, as the need-to-know doctrine still has a role to play. Nonetheless, need-to-know does not exclude the need-to-share philosophy and could easily accommodate it if it is articulated correctly in policy.

Some consideration will need to be given not only to the level at which the analyst's report will be classified but also whether the report (or briefing) will be beneficial to other analysts, commanders, or other agencies that might be working on the same or a related issue. The idea is to discourage those with a mere curiosity but not hinder legitimate access.

Consider this historical case as an example of need-to-know: during the planning of the June 6, 1944, D-Day invasion of Nazi-occupied Europe,

"Serious problems in information sharing also persisted, prior to September 11, between the Intelligence Community and relevant non-Intelligence Community agencies. This included other federal agencies as well as state and local authorities. This lack of communication and collaboration deprived those other entities, as well as the Intelligence Community, of access to potentially valuable information in the 'war' against Bin Ladin."[1]

1. U.S. Congress, Senate and House Select Committees on Intelligence, *Joint Inquiry into Intelligence Community Activities before and after the Terrorist Attacks of September 11, 2001* (Washington, DC: U.S. Government Printing Office, 2002), xvii.

thousands of military commanders were involved. However, only a few, by comparison, knew the full details of the plan. The great bulk of commanders were only privy to the details that allowed them to execute their part of the invasion.

Contrast need-to-know with *right-to-know*. The latter is where a legal precedent exists, allowing a person to have access to information. Usually this precedent is embodied in a statute but may be found in common law. Further, right-to-know may extend to a freedom of information law or a similar law. Analysts should make themselves familiar with laws in their jurisdiction and how they impact what they write and to whom it is disseminated, as unintentional public disclosure may result. Such disclosure may put at risk sources, methods, or information that could jeopardize operations or place people's lives at risk.

HANDLING SENSITIVE INFORMATION

In addressing the issue of security, there are three areas to be examined: (1) personnel, (2) physical, and (3) information. In the defense industry, for instance, standards have long been established to ensure confidential matters are not prematurely disclosed. By way of example, if the Australian navy were to develop a new electronic guidance system for its offensive ship-to-ship missiles, such a project would be expected to necessitate dozens of civilian contractors, numerous military units, and various political leaders.

The logistics of containing the myriad details associated with such a project could be a nightmare if a comprehensive set of classifications and supporting guidelines were not developed. For example, guidelines can be found in the open source literature such as the U.S. Department of Defense's *National Industrial Security Program Operating Manual*[2] and *A Guide to Marking Classified Documents.*[3]

Similarly, in the law enforcement or regulatory context, a set of information classifications and security procedures is mandatory. An example of this is the former Australian Bureau of Criminal Intelligence's guidelines entitled *Document Security.*[4]

Classification of Data

Initial Considerations

The considerations outlined in this chapter are not intended to be rigid in their adaptation but rather flexible in their approach. Readers, especially those in small law enforcement agencies or private sector intelligence firms, must consider many factors before implementing the following countermeasures. Important issues that affect the establishment (or upgrading) of a security program include, but are by no means limited to, financial constraints and the willingness of staff to follow proposed procedures once enacted. There is little sense, for instance, in spending large sums of money on a state-of-the-art, monitored intruder detection system if the costs push a budding private sector intelligence business to the brink of insolvency. Likewise, staff may be tempted to bypass security procedures if those procedures are viewed as overly complicated or time consuming.

Although the security treatments discussed here are from some of the key aspects of good counterintelligence practice, they nonetheless can be adapted either in whole, or in part, depending on circumstances. The important fact is that the principles of counterintelligence are observed and that periodic inspections are carried out to check on the standard of security practiced. For a more full discussion of counterintelligence, see Prunckun's treatment of the topic in *Counterintelligence Theory and Practice.*[5]

Identifying Levels of Threat

Threat identification is the initial step for any intelligence unit when establishing a security program. Likewise, it becomes an ongoing consideration once a security program has been developed. Following are three broadband sources of threat. This list is intended to give the reader an appreciation of the hierarchy of threats that an intelligence unit may face in the course of conducting "business."

Steps taken to thwart intelligence collection at what could be termed a level II threat would be sufficient to guard against any attempt by the inferior level III threat but not the reverse. This is an important factor to remember. It is critical for intelligence units to determine where their threats lie before deciding on the range and depth of security measures they will require. Furthermore, an intelligence unit's threat level may change from time to time due to the dynamics of its operations. Therefore, its security needs will also be required to either escalate or abate in response to these changing conditions.

Level I Threat. Surveillance by a foreign government's security or intelligence agency, or surveillance by one's own national law enforcement/ intelligence organization(s).

Level II Threat. Surveillance by a regional/state law enforcement or intelligence unit, an organized criminal group, a foreign or domestic business competitor employing a "spy-for-hire," a private detective acting on behalf of a party interested in the analyst's unit, or other professional fact finders (e.g., an investigative journalist).

Level III Threat. Nonprofessional surveillance by, say, an employee, a business associate/competitor, or another interested individual or group acting for profit or revenge.

Classification of Information

Business and Private Sector Intelligence

In order to foil possible attempts by a hostile agent to obtain information in the private or business sector, information about the unit's activities should be divided into classifications of sensitivity. Moreover, these classifications should be used as a guide when releasing or disseminating information. (Classification of information in the law enforcement, military, and national security sector is discussed further on in this chapter.) Information in this context means knowledge which requires protection from disclosure in addition to the protection afforded to the intelligence it produces or that is being produced. This includes such areas as:

- Account balances;
- Computer source code;
- Current sales figures;
- Customer/client distribution lists;
- Details of alliances with other intelligence units;
- Information sources;
- Information, technology, and communications networks structures;
- Legal documents/advice;
- New intelligence targets and research projects;
- Operating budgets;

- Patent information;
- Personal details of buyers;
- Personnel (their numbers, positions, salary packages, and expertise);
- Policy directives;
- Production costs;
- Reporting dates and timelines;
- Research methods;
- Research plans;
- Research schedules;
- Sales projections; and
- Strategic and operational plans.

The lowest classification of information designated in the following generic system is grade IV, and it consists of information of a general and unrestricted nature. The type of information provided in company prospectuses is a good example of this. Such information would be suitable for all general inquiries and posting to a website.

The next highest classification, grade III, consists of information which should be available to clients only upon request. Information of this type is best described as information and/or material that, if disclosed to an adversary, could reasonably be expected to cause some degree of "damage" to the intelligence unit.

Moving up the scale again is grade II information. This information should be available to a unit's most important customers. Information with this designation would be information and/or material that, if disclosed inappropriately, could reasonably be expected to cause "serious damage" to the intelligence unit.

Finally, the most sensitive information, grade I, should be available only to staff with a need-to-know and government departments that have appropriate authority. Information of this type, if disclosed to an adversary, would reasonably be expected to cause "exceptionally grave damage" to the intelligence unit. (In a business setting, this classification might mean that the company's bottom line could suffer an impact of five percentage points or more.) Staff authorized to access documentation of this grade should be required to sign a "chain of custody record" in order to assure control over its content. The chain of custody record also facilitates withdrawal and destruction when the documentation is no longer required.

In order to inform staff members of a particular document's degree of sensitivity, each document should be identified with a marking indicating its grade (the roman numerals I to IV in red ink). When marking a document, keep in mind that perhaps not the entire document needs to be classified at a particular sensitivity level. Take for instance a report compiled on a recent research project. It is considered to be ideal for public release in a future counterterrorist awareness campaign in the media (grade IV), but a page (or even several paragraphs) contains technical data about the research that is best kept reserved. That section can carry a grade II stamp, while the remainder of the text displays the general grade IV classification.

By using an information classification system, inappropriate disclosure is less likely to occur, and as the information contained in various documents becomes dated and less sensitive with the passage of time, reclassifying the information's grade downward can then take place.

If sensitive information has been compromised or just "lost," the following guidelines will assist in minimizing the damage that may result:

- Attempt to regain custody of the documents/material;
- Assess the information that has been compromised (or subjected to compromise) to ascertain the potential damage, and institute action necessary to minimize the effects of such damage;
- Investigate to establish the weakness in the security arrangements that caused or permitted the compromise, and alter these arrangements in order to prevent any recurrence; and
- Take actions appropriate to either educate/counsel/discipline the person(s) responsible.

The extent to which an intelligence unit goes to enact a classification system is determined by its size and the overarching authority imposed on it by its mandated creator (i.e., government and military intelligence units will have standards imposed by law or regulation, whereas private sector intelligence units will be guided by policy). In the case of a sole intelligence practitioner or a small research firm, there will be far less need for formal arrangements when compared to large organizations.

SECURITY CLEARANCES

Screening Personnel

Arguably, all agencies hold some form of information that would be valuable to a competitor or an adversary, especially intelligence agencies. If this information is released in an unauthorized manner or prematurely, it could compromise the agency's operations, its personnel, or even the security of a nation.

To mitigate such disclosures, personnel undergo an investigation that leads to a *security clearance*. This type of investigation delves into the employee's character, trustworthiness, and loyalty in order to ensure that he or she can be relied upon to hold secrets secure. The theory is that a person who has been found unreliable in one or more of these areas may be a risk when it comes to guarding information. Or the person may be subject to blackmail and forced to surrender information to avoid disclosure of compromising details of past or present activities.

These investigations, therefore, probe personal character; they examine how people conduct themselves in public and private, test their honesty and their financial position, and inquire about any criminal involvement. They also try to gain an appreciation of their emotional and mental strength as well as other issues that might jeopardize their work with important secrets. Foundational to almost all these investigations are checks of national police records and creditworthiness. In addition, most investigations conduct interviews with individuals who know the applicant's personal behavior patterns and ethical temperament.

Countries may have slightly different security classifications, but generally speaking, national security classifications are divided into categories that comprise the following:

Unclassified. Documents that do not require the protection of a security classification. They may include documents suitable for public release and could include those deemed "for official use only."

Confidential. Information, if disclosed, could reasonably be expected to cause *damage* to national security.

Secret. Information, if disclosed, could reasonably be expected to cause *serious damage* to national security.

Top Secret. Information, if disclosed, could reasonably be expected to cause *grave damage* to national security.

To summarize, a comparison between the illustrative generic classification system discussed above and a national security system is shown in table 21.1.

Table 21.1.   Comparison of the Generic Classification System

Generic Grading System	National Security Example
I	Top Secret
II	Secret
III	Confidential
IV	Unclassified

What necessitates a security clearance? In most cases it is based on one's access to classified information. For instance, if an analyst's job involves accessing confidential information, that person would require a security clearance to the level of confidential. If the person's job required him or her to have access to secret information, they would have to obtain a clearance to secret level.

The screening process for personnel usually starts at the application stage with the applicant completing a detailed *personal history statement* (PHS). This is carried out to prevent both the hiring of unethical people, who may in time disclose confidential information, but also to frustrate any attempt at penetration by a hostile agent. In addition to the applicant's full name, current residential address, and date and place of birth, the PHS can include such subhistories as:

- Marital history;
- Residential history;
- Citizenship history (usually ten years for a clearance to secret level and fifteen years for top secret);
- Educational history;
- Employment history;
- Overseas travel history;
- Military history;
- Criminal history; and
- Details of organizational memberships, as well as character, professional, and credit references.

When reviewing the applicant's PHS, a background investigation will look for any inconsistencies, discrepancies, or unaccountable periods (usually around three months or more). Even though applicants may pass this initial screening process, once they are hired, a probationary period is usually set as a contingency for their possible dismissal should they be suspected of having become a security risk. Similarly, when analysts are promoted or assigned to more sensitive research projects (e.g., top secret), another screening procedure can be conducted, covering the time elapsed from their initial hiring to the present. This is to ascertain if any factors in the analysts' recent past could jeopardize the confidentiality of the information they will be handling.

Another means of safeguarding sensitive information is by drawing up nondisclosure or secrecy agreements. These agreements are intended to create a psychological impression on analysts, reinforcing the importance of protecting information to which they have been entrusted. These agreements are in effect legal contracts and can be used as evidence in a court of law should the analyst be found to be in violation. Nondisclosure

"An investigator must ever realize his tremendous responsibility. On the one hand he is dealing with a most precious commodity—a man's career—and on the other hand he is working to the good of the United States Government. He must conduct his inquiries in such a way as to never do a disservice to either."[2]

2. Harry J. Murphy, Office of Security, Central Intelligence Agency, *Where's What: Sources of Information for Federal Investigators* (New York: Quadrangle/ The New York Times Book Co., 1975), preface.

agreements are also used for temporary research support staff such as data entry operators, computer programmers, and the like.

STORING, PROTECTING, AND DISPOSING OF INTELLIGENCE DATA

Document Storage

An intelligence unit's first line of defense against penetration by a hostile agent is its external barriers—its walls, doors, and windows. Its second line of defense is the "containers" that house its confidential documents, for example, computer servers, desktop workstations, and filing cabinets/ storage shelves. It is, therefore, essential that an intelligence unit identify all documents and records that may be the target of a hostile agent and secure these in containers that minimize the risk of their unauthorized acquisition. The concern is with both the theft of the documents and the undetected theft of the information they contain. So to further reduce the risk of attack on containers designated for sensitive material, the intelligence unit should not store valuables, such as cash, securities, jewels, and precious metals, in them.

Photocopying sensitive documents or digitally photographing them surreptitiously are the two most likely ways a hostile agent could obtain information without a unit's knowledge. Another method, although difficult to attempt, is to remove the document(s) from the unit's offices, copy them, and then return them to their storage container undetected. To guard against the former case, access codes should be installed on the office photocopier to strengthen this potentially weak security link.

To counteract both possibilities, intelligence units should ensure that containers housing their documents have locks and that the locks are used faithfully. Metal containers, such as filing cabinets, with padlocks offer a reasonably high level of security; however, safes and cabinets with combination locks incorporated as part of their physical structure offer a much higher level of protection.

Secure storage is equally applicable to computer disks, portable and USB drives, and tape backups (including offsite storage of backups). The control of keys for an intelligence unit's document containers should follow those guidelines outlined previously.

Document Reproduction

Reproduced classified documents that are confidential, secret, and top secret (or in a private sector intelligence organization—grades I, II, and III) are often marked with the classification of the original material. Only sufficient copies necessary to meet operational requirements are duplicated, and all reproductions are destroyed as soon as they have served their purpose (e.g., concluding a briefing for a decision-making committee). Also, when photocopying sensitive documents, analysts should be cognizant to collect the original(s) before leaving the machine.

Document Safeguards during Use

When confidential documents are not held in secure containers, the analyst using the documents is usually required to:

- Keep the documents under constant visual surveillance;
- Place the documents in a storage container and cover it or turn it face-down when an unauthorized person is present;
- Return the documents to their designated storage container after use; and
- In the case of plans, graphs, charts, or other forms of visual aids, they should be labeled with a code name or code number and not openly bear a designation that could identify the project to an unauthorized observer.

Document Disposal

An intelligence unit's wastepaper basket is an easily accessible source of information for the hostile agent. For instance, it isn't unreasonable to assume that 80 percent or more of the paper generated by a commercial business or professional consulting firm contains information that is confidential to one degree or another. In a national security intelligence agency or a military intelligence unit this figure is likely to be 100 percent. These data are pieces of information that, if acquired by a competitor or

Chapter 21

a hostile opposition, could end in great adversity. This information gathering technique is known in the vernacular as "dumpster diving" and is carried out simply by collecting the day's paper waste before the disposal truck arrives.

For Official Use Only—Despite this label appearing to be a security classification, it is not. The annotation merely alerts the user to the fact that it is protected under a privacy act because it is sensitive data.

An easily overlooked source of information leakage is the photocopier. Spoiled and overrun copies should not be indiscriminately dropped into the wastepaper basket. An important piece of equipment for all intelligence units is a document shredder. These devices are so common now that even retail stores carry them as standard items. An alternative for large units is to use a bulk document destruction service. These companies are usually listed in the Yellow Pages.

Another often overlooked source of information leakage is the impressions left on writing pads. To guard against this, a thin piece of acrylic, plastic, or aluminum should be used under the top sheet of all memo pads and writing tablets to prevent impression marks being left on the pad. Stenographic notes, worksheets, sticky notes, and similar items should be destroyed—not just disposed of. Needless to say, a readable copy can be obtained from any of these sources, and therefore, they are as dangerous as the originals in the hands of a hostile agent. Dictation recorded on tape or digitally should be deleted immediately after being transcribed.

Computer Workstations

Desktop computers pose specific security problems. The chief risks are from unauthorized hardware and software access and software sabotage. In the main, the best countermeasures are those of sound physical and personnel security and software management. Countermeasures designed to protect computer software and data include:

1. Using passwords to authenticate legitimate users of the system. Passwords should be impossible to guess, so it is advisable to avoid using any name that is common or familiar in the work environment, business, or to the project. Also, names that are meaningful to

the user, for example a spouse, child, or pet's name, should be avoided. Passwords consisting of a combination of letters and numbers/symbols are ideal for a very high level of security. Ensure that the software security program that controls user access does not display the password on the screen when logging on and it doesn't appear on any printouts. Users should commit passwords to memory; they should never be posted on terminals, workstations, or notice boards. Above all, users should never tell anyone without proper authority a system's password. A system's password should not be changed at regular intervals but randomly to foil any attempt to anticipate security changes.

2. Isolating information of various grades on separate drives and labeling each with its level of sensitivity, that is, unclassified, official use only, confidential, secret, top secret (or generic grades I, II, III, IV) as outlined in the section on classification of information.

3. Avoiding the use of fixed hard disks for storing sensitive data primarily because they cannot be readily removed for safe storage. A removable/portable hard disk offers a much higher level of security. If a fixed hard drive must be used for work associated with projects involving top secret (grade I) information, the alternative is to store the data on portable or USB drives.

4. Storing all disks (data disks or flash drives, master program disks, and backup program disks) in secure containers.

5. Degaussing damaged or defective drives that contain business information before returning them to the manufacturer or retailer for credit.

6. Overwriting drives before using them for information of a lower classification.

7. Disposing of printouts as outlined in the section on document disposal.

8. Disposing of portable and USB drives by physically destroying them.

9. Shutting down idle terminals.

Countermeasures designed to protect computer hardware include:

1. Bolting the computer base unit to the workstation.

2. Locking the server room when technicians are not in attendance.

3. Positioning computer screens to prevent viewing from windows, doorways, or through glass partitions.

4. Allowing only trusted and qualified technical personnel to service or make modifications to a system.

5. Conducting electronic countermeasure sweeps at irregular intervals for bugs or wiretaps.
6. Shielding cables leaving the server room in metal conduit to prevent electromagnetic radiation, which could be intercepted, and to deter illegal tapping.
7. When disposing of old hard drives, use a commercial disk cleaning software package that writes zeros over the entire disk surface. This will leave the disk usable but sensitive data unrecoverable. If the object is to destroy the disk, then after using a software cleanser, drill four holes (e.g., at approximately 12 o'clock, 3 o'clock, 6 o'clock, and 9 o'clock) through the unit so that each punctures the platter.

And countermeasures designed to guard against software sabotage include:

1. Using only commercial software from recognized, reputable software manufacturers or a custom developer.
2. Loading programs from the manufacturer/designer's original copy or downloading directly from the developer's website.
3. Using only programs on an "approved programs list" in order to reduce the possibility of contracting program viruses from noncommercial software.

   If noncommercial software must be used, for example public domain programs, shareware, freeware, and programs downloaded from the Internet, the software should first be thoroughly examined and tested for the possible presence of the sabotage mentioned above.

   Before screening new, noncommercial, software products for viruses and the like, access to the system's hard disk should be temporarily blocked in order to avoid infection should there be any form of contamination in the program.

   Once new noncommercial software has passed rigorous screening, system users should then be supplied with approved (tested) copies of the program.
4. Implementing or upgrading backup and recovery procedures, which will facilitate a quick and complete reconstruction of a system's programs and data in the event that a saboteur strikes.

Finally, the security measures that an intelligence unit adopts to protect its computer workstations and IT systems should not be discussed with anyone outside of the organization. It is acceptable, however, to acknowledge that measures to combat espionage and sabotage are in place, but the specific techniques and procedures should never be confirmed.

FINAL CONSIDERATIONS

Confidential information should always be regarded as having a finite life span. It must be realized that despite the best-engineered security plans

"The strongest castle walls are not proof against the traitor within."[3]

3. An ancient proverb cited by Richard Clutterbuck, *Terrorism in an Unstable World* (New York: Routledge, 1994), 77.

and the installation of the most sophisticated countermeasures equipment, eventually information that is being guarded will become known to others. The American nuclear fighting capability became known to the entire world on August 6, 1945—the day before it was classified top secret.

Obviously, the best way to keep secrets is to store them in one's head and not communicate to anyone. This is not a very realistic countermeasure in a business environment. The point to be made, however, is that as more people know about some particular secret and the more that is written and recorded about it, the more likely that the secret will become prematurely known to unauthorized people (either inadvertently or by design).

This was the case with the early American atomic bomb research; a Soviet espionage operation was able to penetrate the top secret project and acquire that information well before the rest of the world knew it existed. The second point to be made is that once intelligence agents know or even suspect that someone is guarding secrets, half of their work is done; their next step is to devise a method to acquire it. The question every military, national security agency, and business intelligence unit must consider is: how long is this information going to be secure?

KEY WORDS AND PHRASES

The key words and phrases associated with this chapter are listed below. Demonstrate your understanding of each by writing a short definition or explanation in one or two sentences.

- Classification;
- Compromised information;
- Confidential;
- Counterintelligence;
- Countermeasures;
- Document sensitivity;
- Information security;
- Need-to-know;
- Need-to-share;
- Official use only;
- Personal history statement;
- Premature disclosure;
- Right-to-know;
- Secret;
- Security clearance;
- Top secret; and
- Unclassified.

STUDY QUESTIONS

1. List at least six types of information that would be considered sensitive enough to carry a security classification.
2. List the different levels of security classifications and describe the differences between each.
3. Brainstorm four different categories of hostile agents that your agency might have concerns about.
4. What is the difference among a need-to-know, a right-to-know, and a need-to-share? Give an example of each.

LEARNING ACTIVITY

What does premature disclosure mean? Does this mean that every classified document will be released one day? Research the laws in your jurisdiction as they apply to different types of intelligence documents. Determine which ones are subject to a permanent classification—never to be released—and those that are subject to a time limitation, meaning that one day they will be released to the public.

NOTES

1. For a more detailed discussion of counterintelligence see Hank Prunckun, *Counterintelligence Theory and Practice* (Lanham, MD: Rowman & Littlefield, 2012).

2. U.S. Department of Defense, *DoD 5220.22-M: National Industrial Security Program Operating Manual* (Washington, DC: GPO, February 28, 2006).

3. U.S. Department of Defense, *DoD 5200.1-PH-1: A Guide to Marking Classified Documents* (Washington, DC: GPO, May 2000).

4. Australian Bureau of Criminal Intelligence, *Document Security* (Canberra, Australia: Australian Bureau of Criminal Intelligence, 1987).

5. Hank Prunckun, *Counterintelligence Theory and Practice.*

22

ᔒ

Ethics in Intelligence Research

This topic introduces the notions of accountability and control in collecting data secretly. It also examines the ethical issues involved in presenting research findings by looking at applied social research and how ethics in this field of inquiry compare to secret intelligence research.

INTRODUCTION

Discussions about morality, politicalization, "guideposts," principles, codes, creeds, and values are not the expected reading material for an intelligence analyst. Indeed, such a collection of ethics-based issues is a most unlikely feature of the profession's tradecraft. Yet intelligence professionals face the dilemma of acting ethically while at the same time engaging in what some have portrayed as an unethical business—spying.

> Decision making is not easy, especially when faced with a choice between options that are unappealing or distasteful. However, in intelligence work, this can often be the case.

BACKGROUND

By-and-large, intelligence analysts are recruited with academic backgrounds in such fields as history, sociology, criminology, anthropology, psychology, political science, military sciences, as well as specialized fields such as library science.[1] While obtaining their professional credentials, analysts are likely to have been indoctrinated in the philosophy that they must be ethical—for instance, to take responsibility for the mental, emotional, and physical well-being of those they research.[2]

A number of questions therefore arise, and these concern both research ethics and practice ethics:

- Should intelligence analysts be bound by the same ethical guidelines as their social and behavioral science brethren?
- If so, how does one reconcile being asked to carry out secret intelligence research where the welfare of those researched is not only removed from the fore of the analyst's considerations but also is unlikely to even feature anywhere in the research methodology?
- Likewise, how do analysts restrain their personal opinions from making their way into a formal assessment?
- How do they guard against presenting their own beliefs in the intelligence product?
- How do analysts maintain professional distance and not attempt to influence decision makers through their analytic product yet still provide advice as to options?

Then there are the questions at the opposite end of this ethical issue, exemplified by intelligence assessments on Iraq. It was claimed that intelligence products were created with qualifiers to reflect the ambiguity of the situation in Iraq prior to the 2003 invasion. However, an intelligence analyst then with the Australian Office of National Assessments, Andrew Wilkie, alleged that the Australian government at the time skewed these findings by taking the ambiguity out of the weapons of mass destruction issue and presenting what appeared to be a clear-cut situation to the public.[3,4] Was this the "right" thing to do? How does an analyst deal with this type of problem?

SOME ETHICAL DILEMMAS

Given the breadth and scope of these dilemmas, the question that presents itself is: what guidelines should intelligence professionals follow? As there is no clear-cut answer, the best way to deal with these issues is to read widely and discuss the dilemmas with colleagues and raise these issues at staff meetings.

It would be a good starting point for the analyst to examine the issue of ethics within the intelligence community in its widest context: "'truth' is a goal, yet deception, secrecy and morally troubling compromises are often necessary."[5] Whether it is collecting data by deception, or, say, coercing criminals into becoming informants or people to betray their country, the effects of such actions on intelligence officers can be profound. Having to decide what is "right" can result in self-inflicted psychological damage as well as suffering the real-world consequences from actions taken (reflect on what Wilkie risked by blowing the whistle on the Iraq war and weapons of mass destruction).

As for intelligence collection and analysis, a declassified speech given by the then director of central intelligence, Robert Gates, in the CIA auditorium in March 1992 exemplifies how politicalization manifests itself in different ways—from deliberately distorting analysis and judgments to suit the preferred line of thinking to forcing intelligence products to conform to policy makers' preconceived views. Gates draws attention to the issue of how management can apply pressure to "define and drive certain lines of analysis and substantive view points,"[6] to alter the tone or emphasis of the product or the process that created these products, or to limit alternative viewpoints expressed within.

The points made by Gates were given new currency when they were echoed in the *Report to the President of the United States* by the Commission on the Intelligence Capabilities of the United States Regarding Weapons of Mass Destruction.[7] This report pointed out that the key intelligence briefings to White House personnel and senior executives were skewed. The report said that these intelligence products demonstrated "attention-grabbing headlines and drumbeat of repetition, [that] left an impression of many corroborating reports when in fact there were very few sources. And in other instances, intelligence suggesting the existence of weapons programs was conveyed to senior policymakers, but later information casting doubt upon the validity of that intelligence was not. In ways both subtle and not so subtle, the daily reports seemed to be 'selling' intelligence in order to keep its customers, or at least the First Customer, interested."[8] It's interesting that Gates's insights preceded the commission's findings by some thirteen years—it highlights how short corporate memories can be for "lessons learned."

Nelson Blackstock's book *Cointelpro: The FBI's Secret War on Political Freedom* is an example of how domestic intelligence operations were used to subvert political groups in the 1960s and 1970s.[9] The counterintelligence program ran for decades, ending officially in April 1971. It was argued that the primary purpose was to harass and disrupt legitimate political activity under an alarmist guise of national security (e.g., communists, the New Left, and the Ku Klux Klan).

Putting aside the unpalatable political philosophies that these groups may have advanced, Blackstock presents an equally unpalatable tale of systemic misuse of intelligence by several U.S. government agencies that participated in the counterintelligence activities. As a case study, *Cointel-pro* presents a springboard for ethical discussion by students of the craft of intelligence.

More recent events that could be discussed are the cases of Private First Class Bradley Manning and Mr. Edward Snowden. The former was convicted by a U.S. military court-martial in July 2013 of leaking classified documents, and at the time of this writing, the latter was facing allegations of leaking information about a U.S. classified electronic surveillance program—PRISM. Although both analysts claimed to be "whistleblowers," Manning was convicted of espionage (and other related offenses), and Snowden's defense to his allegations certainly did not establish a clearly supportive case of this assertion.

Although these people claimed to have the good of the nation at heart, one could ask: were the ways they went about exposing (in Snowden's case, allegedly) these perceived injustices the correct way? Were there ways to work within the legislative framework and the rule of law, which underpins the way civil society is governed, to bring about change? Did they cause more harm than they alleviated? These and other points of tension need exploring.

ISSUES WITH INTELLIGENCE DISSEMINATION

Dissemination is the ultimate object of an intelligence research project, but what is it that the analysts intend to do with their research findings? The analysts should have discussed this, but here it needs to be spelled out. Will the results be for publication in an academic journal, circulated to government policy makers, or published in a classified report for operational commanders? Will you formulate any intervention strategies that will form part of your report's conclusions?

Nations can lose sight of their purpose and jeopardize the democratic principles they are trying so diligently to protect. This same issue has been highlighted with regard to foreign policy intelligence. For instance, Roger Hilsman, former professor at Columbia University, succinctly put the case: "Finally there remains the ultimate moral and ethical question, whether the means we use will eventually corrupt our values so as to change the nature of our society just as fundamentally as if we were conquered."[10]

Another key area is that of intelligence operations—that is, covert action—by field operatives. It is important to avoid the alarmist anti-intelligence literature that abounded some years ago and focus on literature that stimulates meaningful debate. James Barry's article "Managing

Covert Action: Guidelines from Just War Theory" is outstanding in this regard.[11]

Barry argues that creating a framework based on "just war" guidelines could prove a credible basis for mounting paramilitary operations as well as those involving various forms of coercion and violence in its different manifestations. Although it goes without saying that it would be unrealistic to think such an approach on its own would eliminate the controversy surrounding covert action, nonetheless, just war theory is a credible platform for announcing the fact that a government is interested in the issue of "right" versus "wrong" and addressing it in an atmosphere of openness and transparency (and in doing so, making sure that this important policy option is still viable).

Finally, there are the ethical issues in some selected parallel professions that have been associated with intelligence gathering—sociology, anthropology, and business. Darren Charters's article on the challenge of ethical competitor intelligence is worthy of note.[12] (Competitor intelligence is sometimes termed *business intelligence* so that its abbreviation—CI—is not mistaken for *counterintelligence*. The abbreviation *CI* is also used by law enforcement agencies when they refer to the protection of *critical infrastructure*.) In his article, Charters provides a method for gauging whether one's actions could be deemed ethical or not. His method—based on an evaluative process—is particularly helpful for analysts who are operating in an environment that does not have formal policies or guidelines in place.

The U.S. *Economic Espionage Act* of 1996 criminalizes the unauthorized appropriation of trade secrets: a) To benefit a foreign government, foreign instrumentality, or foreign agent; or b) For financial or commercial benefit.[1]

1. U.S. Congress, *Title 18 of the United States Code*, part I, chapter 90, 1831–39.

Using the acronym CHIP, Charters constructs a four-factor process by which analysts can weigh the gravity of their proposed actions— community virtues, harm, individual as end, and personal virtues. Using a matrix approach (similar in intent to a SWOT analysis), CHIP compares ethical theories by considering the planned competitor intelligence activity from the perspectives of utilitarian, Kantian, and virtue ethics. Though one could argue CHIP is not a substitute for ethical training, the model does offer a realistic means for benchmarking ethical competitor intelligence activities. Its use would certainly encourage quality and consistency in order to avoid violating professional ethical standards as well as legal statutes—lest we forget the political espionage operation known as Watergate that went wrong.

SOCIETY OF COMPETITIVE INTELLIGENCE PROFESSIONALS: CODE OF ETHICS

- To continually strive to increase the recognition and respect of the profession.
- To comply with all applicable laws, domestic and international.
- To accurately disclose all relevant information, including one's identity and organization, prior to all interviews.
- To avoid conflicts of interest in fulfilling one's duties.
- To provide honest and realistic recommendations and conclusions in the execution of one's duties.
- To promote this code of ethics within one's company, with third-party contractors and within the entire profession.
- To faithfully adhere to and abide by one's company policies, objectives and guidelines.[2]

2. Society of Competitive Intelligence Professionals, *SCIP Code of Ethics for CI Professionals*, http://www.scip.org/About/content.cfm?ItemNumber=578&navItemNumber=504 (accessed May 24, 2012).

Overall, it would be advantageous for analysts to examine the literature on the ethics of intelligence. It might help prompt the formulation of ethical codes for both research analysts and field operatives. Such codes could pave the way for analysts who need to work through the issue of "doing the thing right" when confronted with the dilemma of "doing the right thing."

But, in reality, the distinction between these two paths will no doubt be made with some temperance. Analysts should always be conscious that the final judgment will be grounded in what the people and their constitutionally elected representatives consider "necessary and proper" in the context of the threat (thinking particularly of international terrorism and other transnational crimes). In short, the question needs to be asked: is it an intelligence officer's job to observe and comment, or to warn and protect the nation?[13]

In this regard, intelligence officers need ". . . maturity and wisdom to know what is required for any given situation, and to never be placed in a situation where [the officer loses] sight of the 'big picture' for the narrower questions. This would be equally unethical. So, it is worth repeating—the most unethical position to place yourself, your agency or your nation in, is the position where 'losing" would be judged by history as being as unethical as if you acted unethically."[14]

But who knows what "history" will think if an intelligence officer decides to "observe and comment" rather than "warn and protect," or vice versa. Most decisions will be clouded in uncertainty, especially when reflecting on Hilsman's caution: ". . . whether the means we use will eventually corrupt our values. . . ."[15] In the end, whatever action an intelligence officer decides to take, it will no doubt be made after more than one sleepless night. And when he or she does act on the decision, it may still be deemed unethical, or even illegal.

"A spy must be a man of integrity and yet must be prepared to be a criminal."[3]

3. Bernard Newman, cited in John Alfred Atkins, *The British Spy Novel: Styles in Treachery* (London: John Calder, 1984), 142–43.

Suggestions once put forward by the International Association of Chiefs of Police (IACP) to guard against allegations of unethical conduct by law enforcement field operatives are worthy of note. The IACP advised that in order to protect the rights of individuals genuinely not involved in criminal acts, a set of rules and standards should be adopted to deal with the recording and purging of intelligence files.[16] In the main, these guidelines urge that when an individual's association is not criminal in nature, or criminally related, the information should not be recorded. Likewise, in the case of organizations, unless an organization's "ideology advocates criminal conduct and its members have planned, threatened, attempted or performed such criminal conduct," it is both unnecessary and wrong to gather such information.[17]

Moreover, information about an individual's "sexual, political or religious activities, beliefs or opinions, or any dimension of private life-style" should not be collected or recorded in intelligence files unless that information is material to a criminal investigation.[18,19] In addition, all information collected should be evaluated prior to collation to determine its accuracy. The source also requires evaluation to determine its level of reliability. Information that is not verifiable should not be stored in criminal intelligence files, and likewise, the collection of such information should be strictly limited.

A set of standards is necessary to maintain a successful law enforcement intelligence unit filing system. The elements of such a system include:

1. Specific guidelines should be established for determining:
    a) the kind of information that should be kept in intelligence files;
    b) the method of reviewing the material as to its usefulness and relevance; and
    c) the method of disposing of material purged from intelligence files considered to be no longer useful or relevant.
2. Systematic flow of pertinent and reliable information;
3. Uniform procedures for evaluating and validating information;
4. A system for proper analysis of information;
5. A system capable of rapid and efficient retrieval of all information;
6. Explicit guidelines for disseminating information from the files; and
7. Security procedures.[20]

Systematic purging of files according to such guidelines should ensure that information being collected (as per an intelligence collection plan) is related to approved projects and necessary to meet the decision maker's intelligence requirements.

Situations where data have become irrelevant because of age should not be allowed to develop. For instance, a Rand Corporation study once found an American police department retained intelligence files that "contained information on suspected Nazis, the concern of the 1940s; Communist Party membership rosters, the concern of the 1950s; Black militants, right-wing extremists, and anti-war demonstrators, the concern of the 1960s."[21]

Intelligence managers in the twenty-first century would be hard pressed to justify the retention of these types of data. Analysts should always be mindful of their targets. No matter how unpalatable the cause, they should be conscious of the difference between legal, lawful protest and subversion or criminal activity—distinctions not always correctly made in the past.

For the most part, secrets in the context of this book are secrets of the state (or a private corporation), and as such, keeping them secret will most likely be an obligation of a legal statute or a legally binding agreement. Because of this, and unless disclosure is covered by a countervailing legal instrument (e.g., a whistle-blower-type act, an order of a court, a police warrant, a direction given by a judge or magistrate), then disclosure should not be made. However, if the secret is the cornerstone of a *cover-up* of an illegal activity or it relates to the planning or commission of an illegal activity (or covered under a whistle-blower-type law), then revealing the secret is not likely to fall into this category—in fact, there may be a legal obligation to disclose these facts to a law enforcement or regulatory agency.

By way of example, take the U.S. Army's field manual *Open Source Intelligence*, which lists a number of areas that prohibit the carrying out of intelligence activities, specifically mentioning activities that may be a violation in law. The prohibited intelligence activities identified are the improper collection, retention, or dissemination of U.S. person information:

- Gathering information about U.S. domestic groups not connected with a foreign power or international terrorism.
- Producing and disseminating intelligence threat assessments containing U.S. person information without a clear explanation of the intelligence purpose for which the information was collected.
- Incorporating U.S. person criminal information into an intelligence product without determining if identifying the person is appropriate.
- Storing operations and command traffic about U.S. persons in intelligence files merely because the information was transmitted on a classified system.
- Collecting U.S. person information from open sources without a logical connection to the unit's mission or correlation to a validated collection requirement.
- Identifying a U.S. person by name in an intelligence information report without a requirement to do so.
- Including the identity of a U.S. person in a contact report when that person is not directly involved with the operation.[22]

KEY WORDS AND PHRASES

The key words and phrases associated with this chapter are listed below. Demonstrate your understanding of each by writing a short definition or explanation in one or two sentences.

- Cover-up;
- Democratic principles;
- Ethics;
- Legal obligations;

- Moral dilemmas;
- Politicalization;
- Purging files; and
- Whistle-blower.

STUDY QUESTIONS

1. In your view, should intelligence analysts be bound by the same ethical guidelines as colleagues in the social and behavioral sciences? Discuss the issues that could be considered central to this issue.
2. As an intelligence analyst, cite a practical way(s) you could exercise restraint when it comes to expressing your personal opinions through your written reports.
3. Discuss how analysts might maintain professional distance and not attempt to influence decision makers through their analytic product while still providing advice (hint: the straw man technique for providing a range of policy options).

LEARNING ACTIVITY

Suppose you are asked to carry out a classified intelligence research project on an international terrorist. Discuss how you would reconcile the fact that you are being tasked to carry out secret research where the welfare of the person being researched is not a feature in the project's methodology (in fact, assume her "demise" is the project's objective). In doing so, consider the legal and ethics issues, and weigh up the importance of these factors in your overall judgment.

NOTES

1. Dr. Edna Reid, Federal Bureau of Investigation, Washington, DC, personal communication, June 7, 2011.
2. Gennaro F. Vito, Julie Kunselman, and Richard Tewksbury, *Introduction to Criminal Justice Research Methods: An Applied Approach*, second edition (Springfield, IL: Charles C Thomas, 2008), 45–60.
3. Veteran Intelligence Professionals for Sanity and Andrew Wilkie, "Memorandum: One Person Can Make a Difference," in *The Ethics of Spying*, ed. Jan Goldman (Lanham, MD: Scarecrow Press, 2006), 188.
4. Andrew Wilkie, *Axis of Deceit* (Melbourne, Australia: Pan Macmillan Australia, 2004).

5. Jan Goldman, ed., *The Ethics of Spying* (Lanham, MD: Scarecrow Press, 2006), x.

6. Robert M. Gates, "Guarding against Politicization: A Message to Analysts," in *The Ethics of Spying*, ed. Jan Goldman (Lanham, MD: Scarecrow Press, 2006), 172.

7. Commission on the Intelligence Capabilities of the United States Regarding Weapons of Mass Destruction, *Report to the President of the United States* (Washington, DC: U.S. Independent Agencies and Commissions, March 31, 2005).

8. Commission on the Intelligence Capabilities of the United States Regarding Weapons of Mass Destruction, *Report to the President of the United States*, 14.

9. Nelson Blackstock, *Cointelpro: The FBI's Secret War on Political Freedom* (New York: Vintage Books, 1976).

10. Roger Hilsman, "On Intelligence," *Armed Forces and Society* 8, no. 1 (Fall 1981): 129–43.

11. James A. Barry, "Managing Covert Action: Guidelines from Just War Theory," in *The Ethics of Spying*, ed. Jan Goldman (Lanham, MD: Scarecrow Press, 2006), 248–65.

12. Darren Charters, "Business: The Challenge of Completely Ethical Competitive Intelligence and the 'CHIP' Model," in *The Ethics of Spying*, ed. Jan Goldman (Lanham, MD: Scarecrow Press, 2006), 362–77.

13. George Tenet, former Director of Central Intelligence from 1997 to 2004, paraphrased in Melissa Boyle Mahle, *Denial and Deception: An Insider's View of the CIA from Iran-Contra to 9/11* (New York: Nation Books, 2004), 296.

14. Hank Prunckun, *Counterintelligence Theory and Practice* (Lanham, MD: Scarecrow Press, 2012), chapter 14.

15. Roger Hilsman, "On Intelligence," 129–43.

16. International Association of Chiefs of Police, *Law Enforcement Policy on the Management of Criminal Intelligence* (Gaithersburg, MD: International Association of Chiefs of Police, 1985), 8–9.

17. Los Angeles Police Department, *Standards and Procedures for the Anti-Terrorist Division* (Los Angeles: Los Angeles Police Department, 1984), 2.

18. Los Angeles Police Department, *Standards and Procedures for the Anti-Terrorist Division*, 2.

19. This consideration is applicable to law enforcement intelligence and would not necessarily apply to, say, national security intelligence. In the case of the latter, these types of data may be justified on the basis of compiling a psychoanalytic profile on a target, or for other legitimate reasons. See chapter 10 ("Content Analysis of Qualitative Data") in this book for a discussion of these types of analyses.

20. International Association of Chiefs of Police, *Law Enforcement Policy on the Management of Criminal Intelligence*, 9.

21. Brian Michael Jenkins, Sorrel Wildhorn, and Marvin Lavin, *Intelligence Constraints of the 1970s and Domestic Terrorism: Executive Summary* (Santa Monica, CA: Rand, 1982), 5.

22. U.S. Department of the Army, *FMI 2-22.9: Open Source Intelligence* (Fort Huachuca, AZ: Department of the Army, 2006).

Appendix

֍

Critical Values of Chi-Square Distribution

Degrees of Freedom	$P = .05$	$P = .01$
1	3.84	6.64
2	5.99	9.21
3	7.82	11.35
4	9.49	13.28
5	11.07	15.09
6	12.59	16.81
7	14.07	18.48
8	15.51	20.09
9	16.92	21.67
10	18.31	23.21
11	19.68	24.73
12	21.03	26.22
13	22.36	27.69
14	23.69	29.14
15	25.00	30.58
16	26.30	32.00
17	27.59	33.41
18	28.87	34.81
19	30.14	36.19
20	31.41	37.57
21	32.67	38.93
22	33.92	40.29
23	35.17	41.64
24	36.42	42.98
25	37.65	44.31

Note: This is a facsimile of a table of critical values of chi-square. It has been reproduced here using data that is in the public domain.

Index

6-3-5 method, idea generation, 90
80/20 principle, 325

admiralty rating, 53, *53*, 54, *54*, 55
affinity diagrams, 96–97
agents, 133–135
AK-47, 2
algebraic equations, 269–270
alpha level. *See* statistical significance
al-Qaeda, 28, 269, 286
anatomy of intelligence, 25–34
anthropology, xi, 68, 191, 349, 353
anti-whaling, 7
applied intelligence research, 25–26
ASIS, 20
asset, defined, 292
association analysis. *See* network analysis
attack site analysis, 237–238
attack type analysis, 238
Australian Secret Intelligence Service, 20
average. *See* mean

basic intelligence, 22, 23–24, *27*, 40n6, 194
Big Brother, 4, 119
bivariate analysis, 262–268
black bag operation, 152, 159n11
Black chamber, 20
Blair, Eric Arthur. *See* Orwell, George
Bond, James, 1, 4, *28*
Book of Honor, 2, *3*, 28

Boston, terrorist bombing, 120–121
Bowen, Col. Russell J., 5, 113, 114
brainstorming, 88–91, 92, 97, 175, 184, 186, 325
business intelligence, 9, 12, 36–37, 44, 50, 52, 63, 105, 110, 270, 345, 353
buy-in, 90

Canadian Security Intelligence Service, 20
case study research design, 82
Central Intelligence Agency. *See* CIA
Charles Sturt University, 121
charts. *See* figures
check sheets, 98
chi-square analysis, 262–265, *361*
CIA, 1, 2, 3, 5, 20, 28, 110, 133, 134, 166, 167, 194, 205, 351
clandestine, defined, 131–132
client intelligence. *See* business intelligence
cointelpro, 351, 352, 359n9
Cold War, 11, 28, 110
commodity flow analysis, 199
communications plan, 239–240
competing hypothesis analysis, 179–181
competitive intelligence. *See* business intelligence
competitor intelligence. *See* business intelligence
concept mapping, 93, 95, *96*

363

About the Author

Dr. Hank Prunckun, BS, MSocSc, PhD, is associate professor of intelligence analysis at the Australian Graduate School of Policing and Security, Charles Sturt University, Sydney. He specializes in the study of transnational crime—espionage, terrorism, drugs and arms trafficking, and cyber crime. He is the author of numerous reviews, articles, chapters, and books, including *Intelligence and Private Investigation: Developing Sophisticated Methods for Conducting Inquiries* (Charles C Thomas, 2013); *Counterintelligence Theory and Practice* (Rowman & Littlefield, 2012); *Handbook of Scientific Methods of Inquiry for Intelligence Analysis*, first edition (Scarecrow Press, 2010); *Shadow of Death: An Analytic Bibliography on Political Violence, Terrorism, and Low-Intensity Conflict* (Scarecrow Press, 1995); *Special Access Required: A Practitioner's Guide to Law Enforcement Intelligence Literature* (Scarecrow Press, 1990); and *Information Security: A Practical Handbook on Business Counterintelligence* (Charles C Thomas, 1989). He is the winner of two literature awards and a professional service award from the International Association of Law Enforcement Intelligence Analysts, and is editor-in-chief of *Salus Journal*, associate editor of the *Australian Institute of Professional Intelligence Officers Journal*, and the *International Association of Law Enforcement Intelligence Analysts Journal*. Dr. Prunckun has served in a number of strategic research and tactical intelligence capacities within the criminal justice system during his twenty-eight-year operational career, including almost five years as a senior counterterrorism policy analyst during the Global War on Terror. In addition, he has held a number of operational postings in investigation and security. Dr. Prunckun is also a licensed private investigator.